Advances in Underwater Technology, Ocean Science and Offshore Engineering

Volume 24

Advances in Subsea Pipeline Engineering and Technology

ADVANCES IN UNDERWATER TECHNOLOGY
OCEAN SCIENCE AND OFFSHORE ENGINEERING

Vol. 1. Developments in Diving Technology
Vol. 2. Design and Installation of Subsea Systems
Vol. 3. Offshore Site Investigation
Vol. 4. Evaluation, Comparison and Calibration of Oceanographic Instruments
Vol. 5. Submersible Technology
Vol. 6. Oceanology
Vol. 7. Subsea Control and Data Acquisition
Vol. 8. Exclusive Economic Zones
Vol. 9. Stationing and Stability of Semi-submersibles
Vol. 10. Modular Subsea Production Systems
Vol. 11. Underwater Construction: Development and Potential
Vol. 12. Modelling the Offshore Environment
Vol. 13. Economics of Floating Production Systems
Vol. 14. Submersible Technology: Adapting to Change
Vol. 15. Technology Common to Aero and Marine Engineering
Vol. 16. Oceanology '88
Vol. 17. Energy for Islands
Vol. 18. Disposal of Radioactive Waste in Subsea Sediments
Vol. 19. Diverless and Deepwater Technology
Vol. 20. Subsea International '89: Second Generation Subsea Production Systems
Vol. 21. NDT: Advances in Underwater Inspection Methods
Vol. 22. Subsea Control and Data Acquisition: Technology and Experience
Vol. 23. Subtech '89. Fitness for Purpose
Vol. 24. Advances in Subsea Pipeline Engineering and Technology

Advances in Underwater Technology, Ocean Science and Offshore Engineering

Volume 24

Advances in Subsea Pipeline Engineering and Technology

Papers presented at Aspect '90, a conference organized by the Society for Underwater Technology and held in Aberdeen, Scotland, May 30–31, 1990.

edited by

C. P. Ellinas

Advanced Mechanics & Engineering Ltd.

KLUWER ACADEMIC PUBLISHERS

DORDRECHT / BOSTON / LONDON

ISBN-13:978-94-010-6764-5 e-ISBN-13:978-94-009-0617-4
DOI: 10.1007/978-94-009-0617-4

Published by Kluwer Academic Publishers,
P.O. Box 17, 3300 AA Dordrecht, The Netherlands.

Kluwer Academic Publishers incorporates
the publishing programmes of
D. Reidel, Martinus Nijhoff, Dr W. Junk and MTP Press.

Sold and distributed in the U.S.A. and Canada
by Kluwer Academic Publishers,
101 Philip Drive, Norwell, MA 02061, U.S.A.

In all other countries, sold and distributed
by Kluwer Academic Publishers Group,
P.O. Box 322, 3300 AH Dordrecht, The Netherlands.

Printed on acid-free paper

Conference Planning Committee

Dr P A Frieze
Dr F I Knight
J McGregor Scott
Dr P Raven
Jean Pritchard

Contents

PART III – PIPELINE DESIGN, FABRICATION AND INSTALLATION

Preface

Dr C P Ellinas
Advanced Mechanics & Engineering Ltd

Major advances have been achieved in recent years in subsea pipeline design and installation. Inspection, maintenance and repair have also received much attention. The development of marginal fields has brought with it special problems, which have necessitated novel methods and solutions. In the meanwhile interest in the development of deepwater fields continues with the development of new technology.

This Conference has placed emphasis in addressing developments in pipeline technology under four main headings:

- pipeline/seabed interaction;
- flexible pipelines;
- pipeline design, fabrication and installation;
- deepwater applications.

Advances in North Sea technology over the last few years have been concerned mostly with marginal fields, small diameter pipelines and new materials, which are well covered in the first three topics.

Economic development of marginal fields requires processing of oil and gas to take place not at the wellhead but at existing facilities, usually some distance away. Hydrocarbons are thus often transported at high pressure and temperature in small diameter pipelines, which need to be protected through trenching. However, such operational practice has brought to the fore a problem that in the past was of little concern namely, upheaval buckling. This phenomenon has been the subject of extensive research and novel techniques have been developed to overcome it. Three of the papers presented in this Conference report on the findings of such work and on the successful application of prevention methods. A fourth paper examines the application and behaviour of anchor reinforcement used to stabilise offshore pipelines.

Some of the problems of upheaval buckling can be overcome to a certain extent through the use of flexible pipelines. Their use, however, has increased in recent years in the form of catenary and flexible risers, used in more novel processing installations, such as SWOPS, converted semi-submersibles, etc, and in loading terminals. This increased demand in flexible pipe has been matched by the development and marketing of new products and novel forms of construction. However, one major area which limits the wider application of flexibles, and which still requires further development is their non-destructive testing and the detection of defects and localised damage at an early stage which ensures the initiation of repairs before failure occurs. Work on this and other related subjects is reported in four detailed papers.

Development of marginal fields also often requires that new pipelines are tied into existing facilities, and frequently there is the additional requirement that the lines are piggable. The development, design and implementation of technology which allows pipelines to be tied in while maintaining full piggability has the advantage of eliminating the need for intervention. This is the subject of a paper which also discusses valuable test results.

New materials, higher strength steels, welding and requirements for operation of pipelines at high pressure and temperature are covered in some detail in a number of papers, and facets of these developments are encompassed in the paper reporting on the retrofit installation of risers in Central Brae.

Recent considerations of safety, especially following Piper Alpha, have led to the Department of Energy to initiate a process that will eventually lead to formal safety assessments of offshore installations. These assessments will be required to address installation hardware, human factors, operating practices, procedures and safety management. The implications of these on pipeline design are considered in a wide-ranging review paper, which also discusses major hazards to pipelines and risers in the North Sea. Hazard and protection concepts are examined also with regard to deepwater pipelines. In a session on deepwater applications there is a detailed report on developments in hyperbaric welding technology for pipeline repairs and on techniques and technology developed for pipelaying.

The depressed market in deepwater technology in the North Sea will not last for long, and as the more accessible fields become depleted developments such as covered by papers in this Conference will become central to the economic exploitation of deepwater fields.

The advances in pipeline technology and engineering covered by this Conference, and the technical complexity, diversity and sophistication of the novel concepts and techniques are evidence of exciting progress in what seems to be an inexaustible search for new and more economic ways of fabricating, constructing and maintaining subsea pipelines. As demand grows, it seems that offshore engineers rise successfully to the challenge of responding to an evermore demanding technical and economic climate.

Society for Underwater Technology

The Society was founded in 1966 to promote the further understanding of the underwater environment. It is a multi-disciplinary body with a worldwide membership of scientists and engineers who are active or have a common interest in underwater technology, ocean science and offshore engineering.

Committees

The Society has a number of Committees to study such topics as:

Diving and Submersibles
Offshore Site Investigation and Geotechnics
Environmental Forces and Physical Oceanography
Ocean Mineral and Energy Resources
Subsea Engineering and Operations
Education and Training

Conference and Seminars

An extensive programme is organized to cater for the diverse interests and needs of the membership. An annual programme usually comprises four conferences and a much greater number of one-day seminars plus evening meetings and an occasional visit to a place of technical interest. The Society has organized over 100 seminars in London, Aberdeen and other appropriate centres during the past decade. Attendance at these events is available at significantly reduced levels of registration fees for Members or staff of Corporate Members.

Publications

Proceedings of the more recent conferences have been published in this series of *Advances in Underwater Technology, Ocean Science and Offshore Engineering*. These and other publications produced separately by the Society are available through the Society to members at a reduced cost.

Journal

The Society's quarterly journal *Underwater Technology* caters for the whole spectrum of the inter-disciplinary interests and professional involvement of its readership. It includes papers from authoritative international sources on such subjects as:

Diving Technology and Physiology

Civil Engineering
Submersible Design and Operation
Geology and Geophysics
Subsea Systems
Naval Architecture
Marine Biology and Pollution
Oceanography
Petroleum Exploration and Production
Environmental Data

An Editorial Board has responsibility for ensuring that a high standard of quality and presentation of papers reflects a coherent and balanced coverage of the Society's diverse subject interests; through the Editorial Board, a procedure for assessment of papers is conducted.

Endowment fund

A separate fund has been established to provide tangible incentives to students to acquire knowledge and skills in underwater technology or related aspects of ocean science and offshore engineering. Postgraduate students have been sponsored to study to MSc level and subject to the growth of the fund it is hoped to extend this activity.

Awards

An annual President's Award is presented for a major achievement in underwater technology. In addition there is a series of sponsored annual awards by some Corporate Members for the best contribution to diving operations, oceanography, diverless intervention technology and the best technical paper in the Journal.

FURTHER INFORMATION

If you would like to receive further details, please contact
Society for Underwater Technology, The Memorial Building, 76 Mark Lane, London EC3R 7JN.
Telephone: 01-481 0750; Telex: 886841 1 Mar E G; Fax: 01-481 4001.

Part I

Pipeline/Seabed Interaction

UPHEAVAL BUCKLING MITIGATION BY HOT WATER FLUSHING

I.G. Craig, N.W. Nash and G.A. Oldfield
Sun Oil Britain Limited, Union Terrace, Aberdeen, UK

INTRODUCTION

The Glamis Field in UK North Sea Block 16/21a is operated by Sun Oil Britain Ltd as a subsea step out from the Balmoral Field, 8km to the North East.

The relationship between Balmoral and Glamis is shown in Figure 1. Dedicated single flowlines tie the two production wells and the one water injection well back to the Balmoral template. For process reasons, the production flowlines are heavily insulated to maintain fluid arrival temperatures at the separator above 25°C (Reference 1).

Volume 24: C. P. Ellinas (ed.), Advances in Subsea Pipeline Engineering and Technology, 3–32.

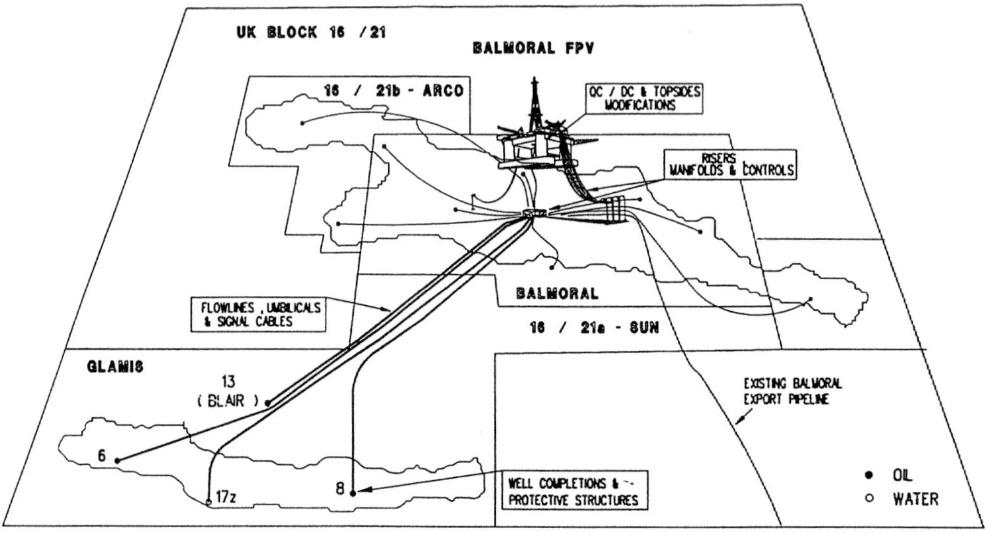

Fig 1: GLAMIS DEVELOPMENT

Pipeline surveys had revealed that an existing Balmoral insulated flowline had been subject to upheaval buckling movements. Similar but more widespread movements were expected on the Glamis lines unless steps were taken to prevent or alleviate the conditions causing the movements . Theoretical predictions based on published analytical techniques confirmed these expectations.

Flowlines buckle when the net compressive load in the pipe exceeds the bending stiffness and the frictional restraining force. Vertical movements occur if the pipe's preferential lateral mode of movement is restrained and the compressive load is able to overcome the weight of the pipe and cover weight/resistance. The compressive load is a combination of the internal pressure effects and the temperature induced expansion forces.

There are several methods for preventing or stabilising buckling movements. Only a limited number were relevant to Glamis and of those only the most appropriate are described in this paper.

The preferred solution was the hot water flushing technique which was developed specifically for this project. The essence of the method is artificially raising the flowline effective installation temperature so that the differential with the operating temperature is reduced to a value which will not cause compressive loads high enough to result in buckling.

Theoretical aspects of the hot water flushing procedure are described in conjunction with details of the offshore operation. The effectiveness of the technique is discussed in terms of the observed pipe movements recorded during the post lay, post hot water flush and post first oil surveys.

SOLUTIONS

Containment or alleviation of upheaval buckling movements can be achieved using either of the two general concepts of cure or prevention. By definition, those techniques based on prevention reduce the compressive loads to below the buckling threshold whereas the cure approach contains the pipe movements when the loads occur.

The important difference between the two approaches is that a well designed prevention method should resolve the issue for the full design life. However a cure method may require regular survey and refurbishment work.

Before discussing the alternative methods considered for Glamis the general conditions under which upheaval buckling can occur are described briefly.

Figure 2 is a diagrammatic presentation of the flowline loads which are caused by the pipe heating up during operation and expanding towards the free ends. At each end of the flowline the load in the pipe is limited by the mobilised soil frictional restraint. Current theories generally assume that once the pipe displacement is greater than about 5-10mm the friction force remains constant as represented by the straight lines in the figure.

Along the central section of the flowline, where sufficient end restraint has been mobilised, the pipe load is governed by the temperature (and pressure) differential between installation and operation. The higher the differential the greater the compressive load. This part of the flowline load diagram is represented in Figure 2 by the line between points A and B. It should be noted that the thermal load line assumes full flowline restraint.

In theory, the load may be calculated directly from the predicted temperature profile. However local lateral pipe movements including a degree of acceptable vertical movement will lower the temperature difference load line appreciably. It should be noted that even though pressure does have an effect on the results it is only marginal and it has been ignored in this discussion.

Superimposed on the flowline load diagram in Figure 2 is the minimum upheaval buckling load line above which unacceptable vertical pipe movements can be expected to occur. For Glamis, unacceptable movements were defined as those that would result in the pipe becoming exposed above the level of the natural seabed.

Close inspection of Figure 2 shows that there are at least four different ways of reducing the Buckling Zone to zero.

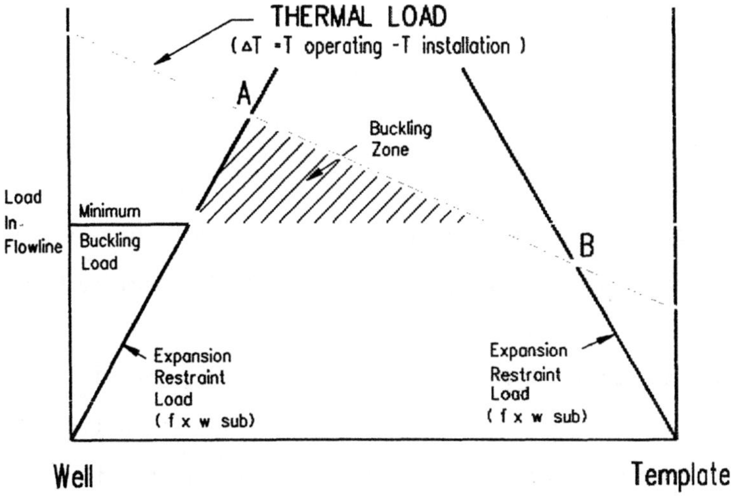

FIG 2 : TYPICAL BUCKLING LOAD DIAGRAM

1. The minimum buckling load line is raised until it exceeds the load represented by point A, e.g. the pipe movement is contained by dumping rock over the line.

2. The temperature difference (thermal) load line is lowered until it is below the minimum buckling load line, e.g. the produced fluid is cooled at the well thereby lowering the maximum operating temperatures.

3. The expansion restraint load lines are manipulated so that they intersect below the minimum buckling load line or where the thermal loads are insufficient to cause buckling e.g. the flowline is divided into shorter lengths with mid line expansion spoolpieces.

4. A combination of two or more of these options.

Prevention Methods

Flowline Cooling

As already indicated in the previous paragraphs the effect of flowline cooling is to lower the thermal load curve below the minimum buckling load line.

To cool the produced fluids down before entering the flowline a number of pipework loops could be mounted within a protective structure and connected in series with the tree and the flowline. The exposed pipework continuously cooled by the surrounding seawater would lead to significant reductions in produced fluid temperatures.

However, low temperatures are often not tolerable for operational reasons. Risk of hydrates, problems with breaking down emulsions in the separators and possible wax deposition in the flowlines and pipework, may require a minimum level of fluid temperatures to be maintained. This was the case for Glamis where a minimum fluid inlet temperature at the separator of 25°C was specified. The option of installing a heater upstream of the separator would not have completely solved the problem.

Mid Line Spoolpieces

Figure 3 shows the effect of introducing mid line spoolpieces. It should be noted however that, for the long Glamis flowlines, more than two flowline sections would have been required.

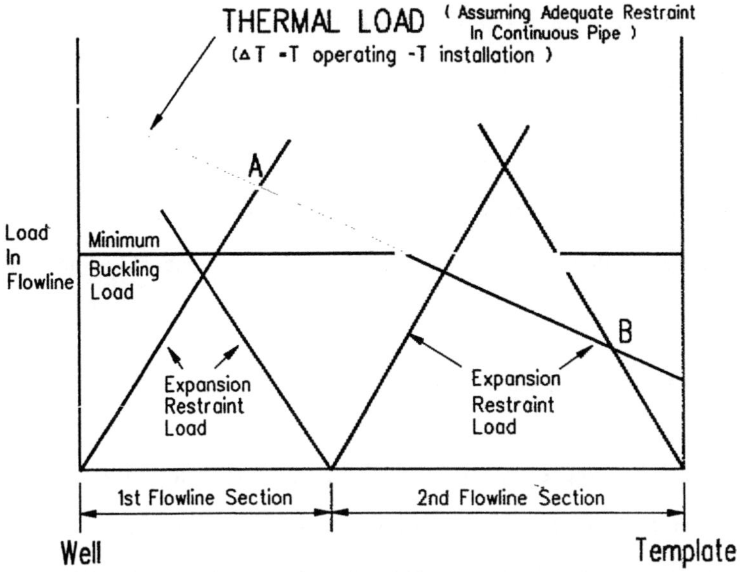

FIG 3 : LOAD DIAGRAM FOR FLOWLINE WITH MID LINE SPOOLPIECE

The spoolpieces are spaced so that at no time does the mobilised soil frictional restraint on the pipe exceed the minimum buckling load. The importance of the effective soil friction factor is evident. To be sure of selecting the optimum spoolpiece spacing a comprehensive soil survey would be necessary.

Although a satisfactory design was feasible there were a number of practical issues which led to this option being dropped.

- Space is required to allow the flowline to expand and the spoolpieces to move. This space could be created by excavating a wide trench or depression in the seabed at right angles to the flowline alignment. Some means of preventing natural backfill to allow unrestricted spool movements is also necessary. Alternative ways of providing space for the spool are feasible but not without increasing costs.

- The flowline installation would take longer with the interruptions to laying. Similarly, the trenching and diving durations would be extended.

- The spoolpieces introduce additional points of potential flowline leakage. Their remote location complicates inspection and maintenance.

Flowline pre-stressing during Installation

Creating a pre-stress in the flowline during installation requires the flowline to be subject to extension (or expansion) displacements which are locked in by soil friction or other mechanical means to prevent the installed pipe from regaining its neutral stress position.

If sufficient tensile pre-stress was applied during installation, it could reduce the operating compressive stress to a level below the minimum buckling stress.

Prestressing the flowline has the same effect as raising the installation temperature and thereby reducing the all important temperature differential between installation and operation. Referring to Figure 2, this is graphically equivalent to the flowline cooling case where the temperature difference load line is reduced so that point A is below the minimum buckling load line.

Initial flowline displacements can be generated by applying a tensile load to the pipe or by heating it so that it expands. Both alternatives have been considered here.

(a) Layspread induced tension

During pipelay the vessel applies tension to the pipe to maintain the pipe catenary and to avoid overstressing of the pipe in the sag bend.

This applied tension will fluctuate during the course of installation both by design, to cater for the different conditions during lay, and by normal operating variations. As the flowline is laid down, full relaxation of this load will be prevented by the mobilised soil friction.

The level of residual tension is difficult to confirm without some means of monitoring the stress in the pipe. However, experience would indicate that in areas like Glamis, where the seabed is a fairly soft clay, the loss in tension would be substantial.

In theory the lay tension could be increased to counter the subsequent level of relaxation. However in practical terms this is not feasible, particularly as the trenching operation, which takes place after the flowline is laid, disturbs the frictional restraint as the trench is formed.

The use of residual lay tension to offset some of the operating compressive load could therefore only be viewed as an unquantifiable benefit and could not be relied on to provide a complete solution.

(b) Hot Water Flushing of the Flowlines.

It is clear that if prestressing is to be effective, it must be maintained during flowline trenching or induced into the line after trenching is complete.

For the Glamis flowlines, consideration was initially given to bringing the well on stream prior to the trenching operation as a way of increasing the effective installation temperature. This solution was discounted because:

- it required DEn dispensation to bring the flowlines into use prior to trenching.
- flowing hydrocarbons while trenching was in progress introduced a significant environmental risk.
- it complicated the normal post trenching gauging operation which is required to confirm no pipe damage.
- hot water flushing could achieve the same results and avoid these risks.

After the flowline is laid on the seabed but before it is trenched, it is heated up by continuous flushing with hot water. The increase in temperature causes the flowline to expand and buckle laterally. The pipe movements will to some extent be restrained by the frictional resistance of the soil. The horizontal imperfections and stiffness of the flowline will govern the proportion of the pipe movement expended in lateral rather than longitudinal displacements.

On completion of trenching the flushing is stopped and the pipe allowed to cool. The contracting pipe is now restrained by soil friction acting in the opposite direction and a tensile stress proportional to the restraint will be induced in the pipe.

The magnitude of the residual tensile stress is directly related to the initial success in maximising the pipe expansion by overcoming soil friction. It is, however, probable that the bulk of the initial resistance not overcome by direct expansion or lateral buckling will be relieved as the pipe is being trenched. If the soil resistance is insufficient to support the compressive loads associated with the initial lateral buckles during hot water flushing it will also not be capable of supporting the higher vertical buckling loads.

After trenching is complete the residual tensile load is dependant on soil friction to sustain it. However, at this stage the pipe is within a trench where the effective soil friction will be higher because of the trench configuration.

Figure 4 presents a graphical comparison of the calculated minimum lateral and upheaval buckling temperatures for a range of friction factors. The analytical method proposed by Hobbs (Ref 2) was found to give results comparable with other techniques and so was used as the basis for all buckling checks. Figure 4 also shows the variation in buckle length for the lateral buckling case as a function of friction factor.

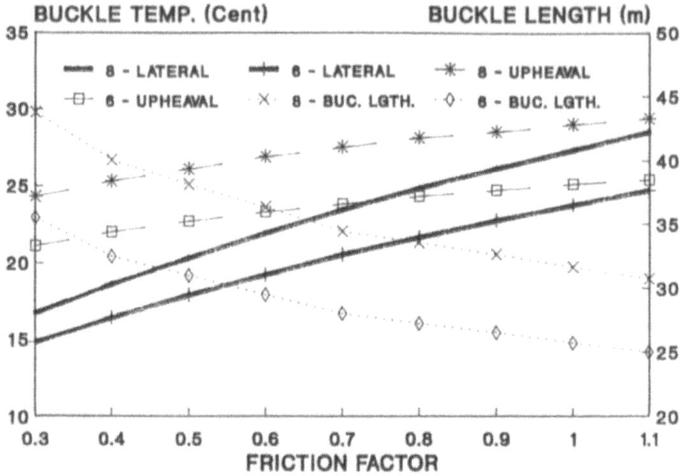

FIG.4- LATERAL V. UPHEAVAL BUCKLING

As the friction factor and therefore the lateral restraint against buckling increases, so the minimum buckling temperatures for lateral and upheaval buckling converge. This convergence is to be expected because it demonstrates that if the loads in the pipe are large enough and there is sufficient lateral restraint the pipe will buckle vertically.

This phenomenon does not mean that the hot water flushing technique is not feasible in high friction factor soils. Any tendency for the flowline to buckle vertically will immediately reduce the soil contact force which is directly proportional to the lateral restraint and lateral buckling will subsequently occur.

During the conceptual design of the hot water flushing technique several conclusions were drawn which were fundamental to the success of the operation.

- Significant lateral buckling will occur at moderate flushing temperatures.

- The low soil restraint required for a lateral buckle to form results in a regular distribution of buckles along the flowline i.e. the bulk of the expansion forces are relieved uniformly along the flowline by lateral movements rather than through longitudinal movement at each end. These movements can readily be locked in by trenching.

- The lateral buckles form horizontal 'imperfections'. At high operating temperatures pipe movements will be concentrated in these areas. These movements will continue to be lateral provided the trench is formed with a sufficiently wide invert.

In conclusion, this technique should prevent upheaval buckling if the flushing water induced tensile load reduces the residual operational compressive load to below the upheaval buckling threshold.

Cure Methods

All methods of containment involve increasing the weight on the pipe or restraining longitudinal movements to restrict associated vertical movements to acceptable levels. This has the effect of raising the minimum buckling load line above point A on Figure 2.

A side effect of this approach is that it steepens the expansion restraint load line WA. This causes the maximum pipe load to increase as A shifts up the curve to the left and increases the tendency of the pipe to buckle.

It is therefore important that if the cure approach is adopted, an adequate factor of safety should be included in the design. The work must also be carried out and maintained effectively.

Flowline Weight Coating

The submerged weight of the flowline can be increased by applying a high density concrete weight coating on the outside of the pipe insulation. The concrete has insulative properties so it would provide a benefit by reducing the insulation specification on the pipe.

A drawback to this approach is that for practical reasons the entire flowline length has to be coated with concrete. However, inspection of Figure 2 shows that the additional weight is only required over part of the line.

Theoretically, a sufficient thickness of concrete could be applied to increase the submerged weight to the specified level. However, the thickness would probably be impractical and uneconomic to lay and trench, particularly as the laying options would be severely restricted.

Deep Burial

If the designed trench depth of 1 metre was increased significantly and the flowline was buried by backfilling, the soil overburden on the pipe could provide the weight necessary to prevent upheaval buckling.

This alternative is very similar to the industry standard rock dumping solution described in the next section. Since the backfill material is the spoil from the trenching operation this method would appear to offer cost advantages against rock dumping which relies on imported materials.

A comprehensive survey of the soil along the flowline route would need to be carried out to assess the variation in properties. The lower density of the reworked spoil would lead to a much greater design trench depth than would be necessary for the rock dump option. Concern about the performance of the reworked spoil resulted in design trench depths in excess of 2 metres. At these depths trenching ceases to be a routine operation. The associated risk was considered too great for this method to be considered further.

Rock Dumping

After the flowline is laid and trenched a rock berm may be dumped along the flowline to provide sufficient vertical load to prevent buckling.

The length and height of the berm required, and its continuity, depends on the estimated value of the soil frictional resistance, the out of straightness of the pipe and the temperature of operation.

The rock overburden provides both local stability against vertical pipe movements as well as increased longitudinal restraint on the flowline, preventing movement towards a buckle zone.

The effect a continuous rock berm, with a mid point change in depth has on the typical load diagram is shown in Figure 5.

The weight of the rock raises the minimum buckling load line so, after allowing for a certain amount of tolerance, the berm is only required over the central flowline section where the pipe loads are greater than the buckling load.

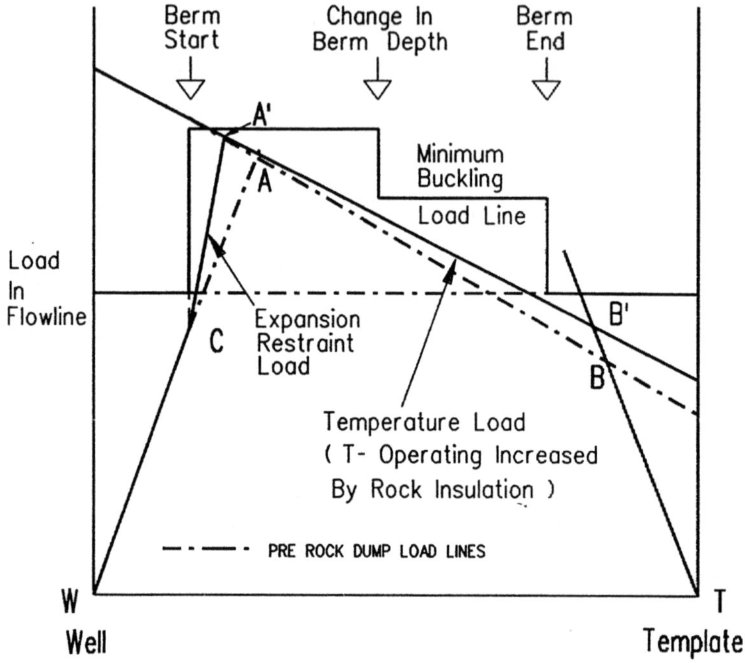

FIG 5 : LOAD DIAGRAM FOR ROCK DUMPED FLOWLINE

The component parts of the load diagram and how they are influenced by the rock is described as follows:-

(i) Expansion Restraint Load

The load is unaffected between W and C which marks the start of the rock berm. Between C and A' the rock weight and friction increase the slope of the line.

(ii) Thermal Load

The insulating effect of the rock raises the operating temperature and therefore the temperature difference relative to installation. This greater load is depicted by the flatter load line between A' and B'.

(iii) Minimum Buckling Load

At the start of the rock berm the load has a step increase which is proportional to the weight of the rock. At the point where the berm is reduced in depth there is a corresponding step down. A similar step occurs at the end of the berm.

Savings on the volume of rock may be achieved by intermittently dumping the flowline along its length with dump lengths and centres related to the minimum distance between buckles. The adoption of this approach is dependant upon a high level of confidence of the operating conditions and the validity of the theoretical analysis.

Rock dumping was retained as the fall back option in the event that the hot water flushing technique proved unsuccessful.

Retrospective Flowline Weight Stabilisation

Evidence to date suggests that if buckling were to occur, it would develop slowly over a period of time. The rate of development has been reported as being related to the number of well start-up and shutdown cycles. The heating and cooling effect encourages the movement of the pipe in towards the location where the buckle is forming.

Therefore, provided there is regular inspection of the flowline, retrospective action can be taken before the buckle has developed to the extent where pipe wall overstressing and crimping is likely to be a problem.

If a buckle does develop, installation of weight stabilisation over the buckle length, to overcome the uplifting force and the loss of frictional resistance over the suspended length of pipe, should be sufficient to prevent further movement.

This approach has a proven track record on the Balmoral flowline where upheaval buckling related movements have been experienced.

It was concluded that retrospective flowline weight stabilisation on the Glamis lines could be used to stabilise any buckles which form if the flowline hot water flushing is not completely successful.

Cost Comparisons

During the conceptual design, capital cost estimates were developed for the contending hot water flush and rock dump options. The estimate for rock dumping was based on a continuous average cover depth of 1.5m over the central section of each line. The hot water flushing costs included the purchase or hire of all flushing equipment, the two heaters, three generators (1 standby), FPV labour, fuel, FPV diving services and a post first oil survey of the flowline.

The estimated cost for rock dumping was £2.0 - £2.5 million. The actual cost for the hot water flushing operation was £0.5m of which approximately £60k was a long lead outlay for manufacture of the hot water heaters.

The magnitude of the cost differential between the two methods could have been reduced by introducing refinements such as intermittent dumping. However it was impossible to trim rock dumping costs to a level where they could compete with the hot water flushing costs.

GLAMIS HOT WATER FLUSHING OPERATION

Conceptual Design

The objective of the hot water flushing operation is to raise the effective installation temperature to a level such that the thermal load line is below the minimum buckling load at all positions along the flowline.

It was clear at the outset that a continuous high volume source of hot water would be required if the installed flowline temperature was to be raised and maintained for a period of days while trenching was in progress.

The availability of such a source on the Balmoral FPV in the form of hot produced water was a key to the viability of the scheme.

In 1988 the average produced water flow rate was approximately 12,000 barrels per day or 300 gallons per minute. With the bulk of the Balmoral production coming from template wells immediately below the FPV, the produced water temperature in the tilted plate separator was around 60°C.

In essence, clean hot water would be drawn from the produced water system and pumped through the flowlines using a typical, pre-commissioning equipment spread.

Several issues still had to be considered:-

(i) To what temperature did the flowline have to be raised during installation to achieve the objective.

(ii) Was the produced water temperature high enough to satisfy (i) or were heaters required.

(iii) If heaters were necessary could the platform supply the power. Alternatively could temporary mobile generators provide the power source.

(iv) The Balmoral FPV is an operating platform with limited available space for large amounts of temporary equipment. Could space be allocated for up to four weeks for this operation, particularly if heaters and generators were required.

(v) The specifications for the standard flowline flushing pumps, flexible hoses and ancillary equipment had to be reviewed for compatibility with the hot produced water.

Flushing Water Temperature

A number of theoretical models have been published for predicting minimum upheaval buckling temperatures. Preliminary calculations were carried out using a modified version of the method proposed by Hobbs (Reference 2).

These calculations indicated that for both Glamis flowlines buckling would occur at operating temperatures in the low to mid twenties centigrade, depending on the soil friction factor.

These values appeared to be very low when compared with existing flowlines which were operating at higher temperatures and yet had shown little or no sign of any movement.

Sun Oil Britain have experienced a case of upheaval buckling movement on the production flowline from the most distant Balmoral satellite well. Initial attempts to correlate the theory with observations of the flowline movement proved unsuccessful.

Clearly if the theory was correct some other factor or factors were effecting the result. After a detailed review of the flowline as built and post buckle video surveys, it was concluded that significant lateral displacements were probably occurring within the flat invert of the trench profile. These movements were allowing a partial relaxation of the compressive load. Such movements could not be allowed for in the theoretical analysis of the flowline. In addition, the simplistic elastic analysis could not simulate the small elastic/plastic rotations which occur at higher stress levels.

At the point where the buckle occurred, the flowline post trench position had been tight against the trench wall on the outside of a curve. This prevented significant sideways displacements with the result that most of the pipe movement was vertical. Once the pipe moved above the top of the trench and lateral restraint was removed, the flowline reverted to the lateral buckling mode.

The foregoing explanation suggests that in the graphical presentation of Figure 2 the buckling zone is exaggerated by underestimating the minimum buckling load and overestimating the thermal load.

A trench profile with a flat invert would obviously reduce the thermal load but the scale of the reduction is difficult to quantify. It was therefore considered appropriate to calculate the thermal load line assuming no relaxation of load but to raise the minimum buckling load line to reflect these lateral movements.

Based on the Balmoral data, and allowing for the heavier and stiffer Glamis pipe sections, the minimum buckling load for the Glamis flowlines was conservatively calculated as being equivalent to a 58°C (±9°C) thermal load.

Referring to Figure 2, the minimum hot water flushing temperature was equal to the flowline maximum operating temperature at A less 58°C. To fix point A, assumptions had to be made concerning the build up in soil restraint on the flowline. To avoid doing so and provide a margin for error, A was conservatively assumed to be at the wellhead. "T operating" was therefore equal to the maximum flowline inlet temperature of 103°C, i.e. the design hot water flushing temperature at the wellhead end of the flowline was 45°C (±9°C).

A detailed explanation of how these temperatures were derived is given in Reference 3.

Electric Heaters and Power Generation

To achieve the calculated flowline discharge temperature the required flowline inlet and platform flushing pump discharge temperatures were assessed. The flushing flow rate obviously has a large impact on these temperatures. A design flow rate of 200 gpm with an absolute maximum of 250 gpm was selected to allow a margin of at least 50 gpm to be discharged into the sea in the normal way.

The design envelope for the 16/21a-6 flowline is shown in Figure 6. It soon became apparent that the temperature of the produced water was insufficient to satisfy the design parameters and heat exchangers and generators would be required to boost the flushing water temperature. Since the temperature drop between the heater and the flowline inlet was approximately 5°C, the heater design specification was to heat the produced water at the rate of 200 gpm from an inlet temperature of 50°C (assumed for design purposes) to an outlet temperature of 75°C.

Figure 6 also shows the inferred temperature profiles along the flowline based on recorded inlet and outlet temperatures for the hot water flushing operation and projected temperatures for the operating condition.

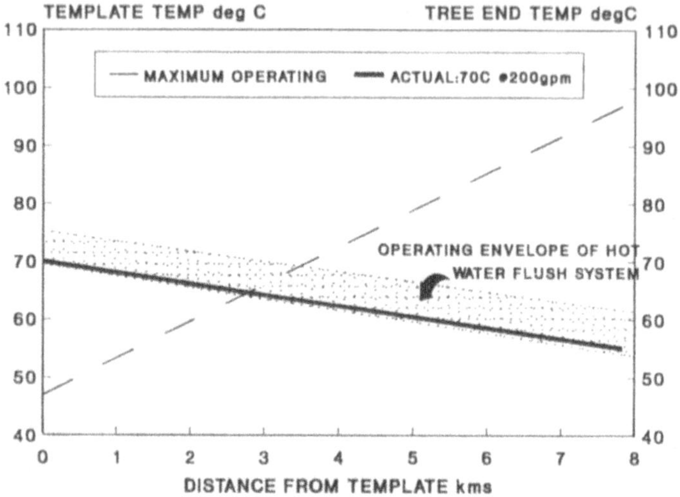

FIG 6- 16/21a-6 FLOWLINE TEMPERATURES

To provide security against breakdown at a critical stage of the operation a decision was made to provide two 100 percent flushing trains in parallel. Isolation of one train would permit continued operation of the other. To satisfy this criterion two 1 megawatt electric heaters powered by two 1500 kVA generators were required.

System Description

Figure 7 provides a generalised schematic of the hot water flushing system.

The flowlines were laid with a pig launcher in the initiation head and a pig receiver in the laydown head. A temperature probe with an integrated acoustic telemetry system was installed with the laydown head which permitted remote monitoring of the flushing water discharge temperatures.

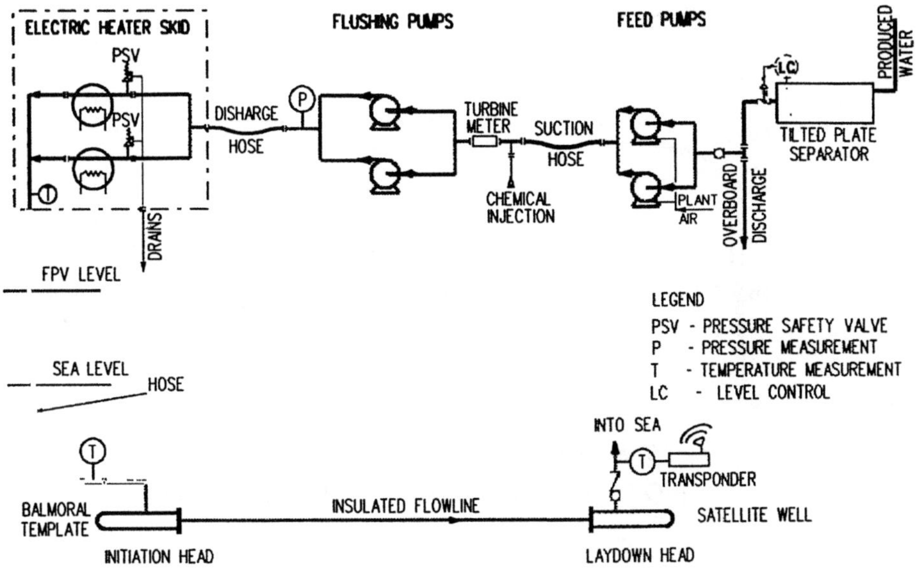

FIGURE 7 - FLOWLINE HOT WATER FLUSHING SYSTEM SCHEMATIC

On the FPV the produced water was drawn off the discharge drain via a temporary tee. Two air driven booster pumps maintained a positive suction pressure at the flushing pumps, which pumped the water through the heaters and down into the flowline via a flexible hose suspended through the vessel moonpool. A temperature probe in the initiation head recorded flowline inlet temperatures.

To prevent scaling from the produced water a scale inhibitor was injected into the water upstream of the flushing pumps.

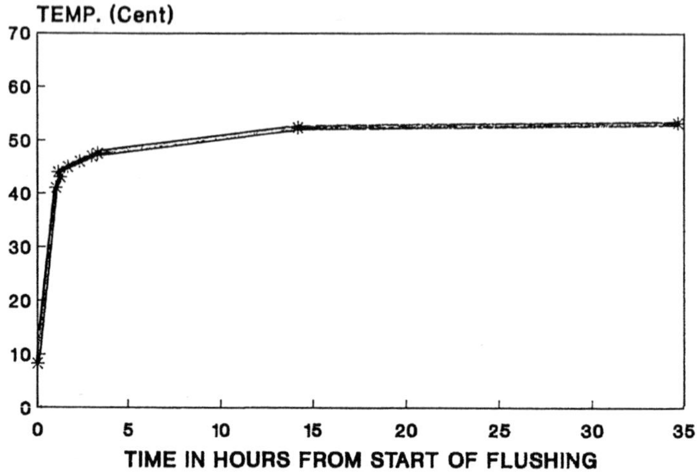

FIG.8- 16/21a-6 OUTLET TEMP. V. TIME

Flushing Procedure

After the flowlines had been laid the flexible flushing hose was connected to the initiation head of the first line and flushing commenced.

A time lag of approximately 12-18 hours had been calculated between start up and the pipe reaching the specified temperature. In fact this proved to be a conservative estimate, as evidenced by the discharge temperature versus time curve for the well 16/21a-6 flowline presented in Figure 8.

With the flowline up to temperature the survey vessel carried out a detailed side scan and ROV visual survey of the post hot water flush flowline to establish the extent of the lateral buckling. Following receipt of the recorded changes to the flowline alignment, the trenching vessel deployed the plough onto the seabed astride the pipe and commenced trenching.

During concept evaluation it had been predicted that there could be some movement of the pipe ahead of the plough as it systematically removed the frictional restraint on the pipe in the process of forming the trench. This effect was likely to be accentuated in the lateral buckle regions. This concern proved to be unfounded. Whatever movement occurred was masked by the normal action of the plough as the pipe was picked up, passed through the rollers and set down above the newly cut trench.

On completion of ploughing the flushing was stopped and the flowline allowed to cool. A post trench survey was carried out to check the position of the pipe and confirm the target depths had been achieved.

FLOWLINE SURVEYS

General

Both flowlines were surveyed after laying, after hot water flushing temperatures had stabilised, after trenching and some weeks after first oil.

To simulate the highest operating temperatures which occur later in the field life, the production rate from well 16/21a-8 was temporarily boosted by opening up the choke. With the flowline stabilised at the higher temperatures a further survey was carried out.

Survey Results

Although it was not possible to check the temperature at which lateral buckling had occurred, the geometry and buckle intervals were very similar to the theoretical predictions.

The accuracy of the surface navigation system was insufficient to be able to quantify the relative pipe movements between surveys. However the surface veneer of soft sediments in the 16/21a block left tell tale scuff marks where the pipe moved sideways.

By close inspection of the continuous video surveys it was possible to estimate relative movements of the pipe.

It was difficult to accurately compare the pre and post trench surveys because the pipe movements were small and tended to be masked by the movements caused by the plough.

The maximum reduction in lateral buckle amplitude as a result of the pipe cooling after shutting down the flushing pumps was estimated at one pipe diameter. In some areas the pipe hardly moved.

The post first oil survey was carried out approximately five weeks after production commenced. In spite of the time interval it was still clear from the disposition of the sediment adjacent to the pipe that the maximum increase in lateral buckle amplitude was of the order of three pipe diameters. Almost all pipe movements were amplitude increments at the identified lateral buckle locations. The two or three areas of lateral movement not in this category were associated with previous buckling which had been too small to detect. There was no evidence of any vertical buckling except in one area where the laterally buckling pipe had contacted the side wall of the trench and moved up it about 300mm.

The uplifted post first oil survey carried out on the 16/21a-8 flowline showed no change in most of the buckle areas. Further lateral movements of up to one pipe diameter were seen but in general the uplifted production rates did not have a significant effect.

CONCLUSIONS

The hot water flushing operation on the Glamis flowlines to prevent upheaval buckling was successful. The procedure artificially raised the flowline installation temperatures to levels which reduced the subsequent operating thermal movements to acceptable levels. These additional pipe movements were with one exception contained within the purposely designed flat invert of the trench.

The experience gained during implementation and execution of the work has confirmed that the technique is a viable and cost effective way of preventing upheaval buckling in circumstances where there is an adequate supply of hot water and/or power generating capacity.

The number and magnitude of the observed flowline movements leave little doubt that upheaval buckling would have occurred had the hot water flushing operation not been carried out and the trench profiled to allow further lateral movement.

As predicted, the lateral buckles formed horizontal imperfections at which the majority of all subsequent movements were concentrated.

The independence and flexibility of the hot water flushing technique avoided the schedule extensions and delays which rock dumping would have introduced. The operation was not without its difficulties but these were typical of offshore construction work. The relative smoothness of execution was due in large part to the use of standard equipment with experienced offshore contractors, supported by the resources and commitment of the Balmoral FPV operations staff.

ACKNOWLEDGEMENTS

The authors would like to thank Sun Oil Britain Limited management and its partners in the P201 Group [Deminex UK (Balmoral) Limited, Clyde Expro plc and Lasmo (TNS) Limited] for the opportunity to publish this paper. The authors also acknowledge the support and assistance of Sun staff and contractors in the preparation of the paper.

REFERENCES

1. Craig, I.G. and Oldfield, G.A.: "Glamis: A Subsea Satellite Field Development", OTC 6201, 1990.

2. Hobbs, R.E.: "In-Service Buckling of Heated Pipelines" Journal of Transportation Engineering, Vol. 110, No. 2 (March 1984).

3. Craig, I.G., Nash N. and Oldfield, G.A.: "Upheaval Buckling - A Practical Solution Using Hot Water Flushing Technique", OTC 6334, 1990.

PIPELINE DEBURIAL

N. V. Sills

Consortium Resource Management Ltd.

19 Denmore Ind. Est., Aberdeen

ABSTRACT

The paper gives a brief history of pipeline burial in the North Sea to establish the parameters for deburial and indicates why this activity is increasing.
A review of past and present deburial techniques and equipment is given together with their most important features.
The paper concludes with a look at future developments.

Volume 24: C. P. Ellinas (ed.), Advances in Subsea Pipeline Engineering and Technology, 33–50.

BACKGROUND

Since the first North Sea gas export pipeline was laid some 21 years ago, oil and gas companies have been obliged to bury their pipelines for reasons of stability in strong currents, protection from impact damage, thermal insulation and to meet governmental requirements.

The earliest pipeline burials were carried out in the mobile sand and shell soils of the shallow Southern North Sea. Any visitor to the Southern Sector in the late sixties and early seventies will recall the familiar pipelay barge followed several miles later by the unmistakable profile of a jet barge with its'pipe burying "claw"

These units, when towed along the pipe, blasted huge volumes of high pressure water below and to the sides of the pipe and disgorged fluidised sand through large airlifts either side of the pipe. The effect was to half trench, half fluidise the ground and allow the pipeline to settle 1 to 2 metres below the natural seabed. The weight of the "claw" often assisted by pressing the pipe into the fluidised soil. This work was normally only undertaken on unpressurised pipelines.

Later more sophisticated low pressure soil fluidisation machines were developed, like that of the Netherlands Offshore Company that fluidised large areas of seabed and allowed the pipeline to sink under its own weight. Some pipelines were thus buried even more deeply, up to 3 metres on occasions.

Subsequent movement of large sand waves across these areas resulted in pipelines being covered by a total of 5 metres or more of soil. At other times subsea conditions reversed this trend and left pipelines exposed on the surface and the battle was on to backfill scours, support freespans and rebury exposed sections.

At a time when the objectives were always to lay, bury and test pipelines to bring fields on stream as fast as possible and to keep pipelines buried, there was little incentive to develop methods of reversing the process or to dig up live pipelines.

When oil and gas production moved northwards into the much deeper Northern North Sea different seabed materials were encountered, silts, silty clays and boulder clays.

These soils with shear strengths often between 100 and 200 KPA (eg Fulmar 130-160 KPA) but sometimes exceeding 400 KPA and resembling tough, hard plasticine, could not be trenched with the traditional jet barge and demanded a different approach to burial.

A new generation of remotely controlled caterpillar tracked trenching machines and towed ploughs evolved which employed mechanical digging devices and furrow forming blades much more suited to these cohesive soils. Both pre and post trenching pipelay methods were used as trenches cut in these high strength soils often stayed open for many months, sometimes even years before the slow collapse of trench walls and sedimentation backfilled them - the reverse of the problems encountered in sandy soils.
Due to the stability of the soil, operators could accept less depth of burial and the top of the pipelines were usually only 0.5 metres or so below natural seabed.

In other North Sea areas in very deep water or where very high soil strength made trenching uneconomic, operators elected to stabilise and protect pipelines by rock dumping. Most often this took the form of a 0.5 to 1 metre thick layer of 75 to 150 mm diameter natural rock. In some areas spent ores and industrial slag have been used whose interlocking shape and tendency to slowly cement together may make deburial difficult.

Small diameter infield pipelines and flowlines have largely been left unburied or sometimes partially covered with protective mats, covers or aggregate.

The close proximity of some pipelines and infield services to drilling operations, such as those next to multiwell subsea templates and production platforms has lead to a surprising degree of unintentional burial. Several operators now have lengths of pipeline submerged in cuttings mounds 10 metres deep. In some fields the drift of cuttings from 40 wells (some 20,000m3 of spoil) has covered pipelines with a metre depth of material as far as 50 metres out from the platform. Any work on these facilities would require some degree of deburial.

THE REQUIREMENT FOR DEBURIAL IS INCREASING

Advances in technology, such as the development of enhanced oil and gas recovery techniques and the discovery of new reservoirs has lead operators to seek ways of extending the life of pipelines over their original design life.

To ensure environmental and economic safety, periodic pipeline inspections have to be made and maintenance work undertaken. Despite recent advances in internal pipeline inspection the high cost and disruptive effect on production has lead some operators to prefer deburial and external inspection.

By-passing redundant structures and depleted reservoirs and bringing on stream new fields may require rerouting of existing pipelines and tying-in new ones. Accidents such as the tragedy of Piper A and the ferry sinking on a Danish pipeline have lead to considerable deburial activity to effect repairs.

Measures to improve the safety of platforms and pipelines such as the installation of subsea emergency shut down valves will require substantial deburial activity in the future.

The decommissioning and removal of redundant pipelines such as that undertaken for Wintershall in the Dutch K15 sector show that it is no easy matter to extract even a small pipeline once settled into the seabed. The embrittlement of the pipeline material and the effects of corrosion, erosion and fatigue may not allow the deeper pipelines to be simply pulled out of the seabed without continual buckling and breaking. Prior deburial may be required to reduce loading.

IMPORTANT FACTORS IN DEBURIAL

Just as the methodology of burying a pipeline is heavily influenced by the soil characteristics so is that of deburial. Deburial is not a simple reversal of the burial technique as the parameters are different. A pipeline will not float neatly back to seabed level when uncovered. Burial of a 1 metre diameter pipeline in 2 metres of mobile sand by fluidisation simply required the pipeline to sink under its own weight into the ground. No trench was actually made. To debury the same, probably live, pipeline would require removing the 2 metre cover and an additional 1 metre - the diameter of the pipeline. To access the bottom of the pipe may require removing a further 0.5 metre a total trench depth of 3.5 metres.

In mobile sand such as that common in Southern North Sea areas the stable angle of repose of the soil is about 15 degrees at best and 7 degrees at worst.

To dig a 3.5 metre deep trench in this soil with a 15 degree angle of repose requires a trench width of 26 metres At an angle of 7 degrees the trench width increases to over 60 metres, about the width of a motorway. A typical trench (10 degree slope) would be 40 metres wide. Each 1 metre length of trench would require the removal of 70 cubic metres of soil. The spoil from deburying a 1 kilometre length of pipeline (70,000m3) would cover an entire football pitch in 7 metres depth of sand and have a dry weight of some 200,000 tons.

At the other extreme where pipelines were laid in trenches cut through hard or highly cohesive soils, such as boulder clay or homogeneous clays different factors come into play. Many trenches dug ten years ago are still visible as shallow depressions in the seabed. The slow movement of bottom currents and slow sedimentation rate make natural backfilling a long process.

In some cases the original trenches, perhaps some 1.5 to 2 metres deep and 3 to 4 metres wide may have slowly collapsed and assisted burial.

When it comes to deburial this "disturbed'" backfill can be very much weaker than the original undisturbed material still forming the trench walls. It will probably have quite different characteristics. These characteristics depend to a large extent on the type of trenching machine used to dig the trench. Machines using chain carried "scoops" or rotary cutting teeth and high pressure water jets will break up the soil almost entirely, depositing it along the banks of the trench. The spoil will slowly return to the trench to form a soft relatively homogeneous mud which even after 10 years is unlikely to exceed one quarter of the original soil strength. Where ploughs have been used the soil will be much less disturbed and if spoil has been returned to the trench by a backfill blade the spoil strength may be much nearer its undisturbed shear strength and probably lumpy.

Because of the much higher angles of repose of these soils (up to 90 degrees) when compared to sandy soils (say 10 degrees) only about one fifth the volume of soil needs removing to achieve the same trench depth.

However, where in sand, to gain a 0.5 metre clearance beneath a pipeline requires a substantial increase in trench width, in cohesive soils it will mean excavating downward in high shear strength "virgin soil" - a completely different problem. Equally to position pipehandling frames or welding habitats will involve expanding the excavations sideways into virgin seabed and presents different problems in different soils.

REQUIREMENTS FOR DEBURIAL SYSTEMS

Apart from the removal of redundant, decommissioned pipelines and accident repairs, deburial will most often be undertaken on live pipelines. Only after deburial has been completed and other preparatory work progressed is an operator normally willing to "turn off the tap" and incur the expense of lost production and perhaps the prospect of servicing shut-in wells. For inspection and maintenance work not requiring depressurisation an operator is unlikely to favour a deburial system that forces him to do so.

The first requirement must therefore be pipeline "friendliness"

In todays economic environment the second requirement would normally be cost effective performance which with the high cost of offshore support vessels invariably demands a rapid excavation rate.

Other requirements, such as sensitivity to weather conditions, tool size and accessibility in confined areas, availability and in areas where soil conditions are not well known, versatility to perform in a range of soil types.

In only very restricted circumstances can the system used to bury a pipeline be used to debury it. The large and heavy tracked systems such as Heerema's Eager Beaver and Allseas Digging Donald that follow a pipeline laid on the surface, excavate a trench underneath it and moved on allowing the line to settle into the trench behind it, cannot reverse the process but only dig deeper.

Likewise the vessel towed trenched ploughs like those of Technomare and Smit are unsuitable for deburial. With bollard pull requirements sometimes over 50 tons misalignment on the pipeline could cause damage.

Neither of these methods can be used for deburial in deep sandy soils.

Of more significance are the large track mounted jet pumps such as Alluvial Mining Company's "Tramrod" and more recently Consortium Resource Management's Remote Underwater Excavator "RUE" both of which are designed for deburial in a wide range of soil types and at high rates of excavation.

The recent upsurge in deburial activity has stimulated excavation companies to further develop their techniques and equipment. This fact is borne out by a review of deburial techniques.

REVIEW OF PIPELINE DEBURIAL SYSTEMS

In the 1970's little requirement existed for deburying pipelines. Most of the work was confined to excavation in shallow water and sandy soils. Small scale local excavations were required to enable divers to work on pipeline flanges and install spool pieces close to platforms. The only available tools were medium pressure diver operated water jetting tools that consisted of a 25 millimetre diameter "T" piece supplied with water through a fire hose. Small excavations 1 to 2 metres deep could be accomplished at rates of 1 to 2 cubic metres per hour.

Airlifts were also common and consisted of a 5-10 metre long open ended tube supplied at the intake with compressed air. The maximum size divers could handle, especially in strong Southern North Sea currents was about 250 millimetres in diameter. This size of airlift could dredge sand, gravel and weak muds at rates up to 5 cubic metres per hour but had to be operated in a near vertical attitude and became very inefficient at depths greater than 75 metres. Additionally they relied on strong currents to carry away spoil.

In the late 1970's and early 80's diving contractors working in deeper waters experimented with water lifts in an effort to improve dredging rates and to allow dredged materials to be pumped horizontally. These early tools tended to be 75-100 millimetres diameter tubes with a single central high pressure water jet forming a venturi. Some had short exhaust hoses. Their performance was generally poor, less than 1 cubic metre per hour and they were subject to frequent blockage.

As the requirement for underwater dredging increased, more sophisticated diver operated water powered jet pumps were developed that operated on medium pressure water (10 BAR) and had less restricted intakes. An annulus shaped water jet replaced/

replaced the central water jet. They were much more powerful and could be equipped with much longer exhaust hoses to carry dredged material, 50 metres or so from the dredge head when assisted by boost pumps. Later versions incorporated water jets or mechanical cutting heads to improve performance in higher strength soils. Production rates for these 100 millimetre diameter units ranged from 10 cubic metres per hour in free flowing sand to below 0.5 cubic metres per hour in strongly cohesive soils (greater than about 75 KPA). As with all diver operated tools the expertise of the individual using them will influence productivity.

In the mid 1980's marinised centrifugal pumps such as the Toyo pump were introduced capable of dredging higher strength soil types. They incorporated a mechanical agitator in the dredge head to assist production but the weight and size of the dredge head (approx. 1 ton) required a lifting frame or buoyancy aid to manoeuvre it underwater. They are therefore difficult to use in restricted spaces. A 150 millimetre exhaust hose fitted to the pump carries slurry over considerable distances and production rates vary between about 15 cubic metres in sand to 2 cubic metres per hour in cohesive soils.

In recent years large track mounted jet pumps such as British Telecom's Seadog and Alluvial Mining Company's Tramrod have been introduced capable of much higher excavation rates. The biggest (Tramrod) has a single boom mounted 250 millimetre diameter jet pump and can achieve excavation rates of 200 cubic metres per hour in sand and upwards of 25 cubic metres per hour in silty clays. Its size, weight (15 tons) and large discharge hose has restricted its use close to live pipeline systems and in areas with restricted access.

The much smaller and more mobile ROV mounted 100 millimetre jet pumps have some notable successes in shallow excavations and removal of fluid drill cuttings but their low excavation rate, generally less than their more powerful diver operated counterparts make them unsuitable for deburying large diameter pipelines.

Perhaps the biggest operational drawback with suction dredging systems is the accuracy with which the suction head has to be positioned relative to the work material, typically within 50-150 millimetres. The smaller units have a tendency to blockage by rocks, shells and debris.

In more recent years an entirely different principle of excavation, that of using low velocity (12 knots) large diameter water columns to mobilise and wash away seabed materials, has been used in Consortium Resource Management's "RUE" systems. Its high excavation rate ranging from 200 cubic metres per hour in sand to 20 cubic metres per hour in cohesive silty clay (100 KPA), relatively light weight (2.7 tons) and buoyant mode of deployment underwater has made it suitable for deburying live pipelines. Like Tramrod and Seadog it is capable of operating in 300 metres water depth but requires very little support in terms of specialised handling systems or deck space. An additional advantage is its ability to automatically adjust its operating altitude (normally some 1.5 metres above the work face) and to achieve extensive excavations from one position.

FUTURE DEVELOPMENT IN PIPELINE DEBURIAL

In considering the future development of deburial systems the first question to ask is, what must the ideal system be capable of doing? The answer is probably, debury all sizes of pipeline from any depth of cover and in any soil strength. However/

However a more realistic answer is to set parameters that would satisfy the majority of operators requirements and least compromise equipment design. The author would welcome comment from conference delegates regarding their own company's requirements.

For commercial reasons the several companies engaged in developing and building new systems cannot openly discuss their designs. However, one company's view and that of the author is embodied in a development programme presently underway and funded by the EEC and major oil and gas companies. The stimulus for the programme is that market forecasts for 1990 pipeline deburial predict a two fold increase in activity over the 1989 level and further increases in 1991 & 92.

The aim is to build self-contained remotely operated excavation systems purpose designed for pipeline deburial that can be deployed prior to the main work vessels arrival. The configuration and light weight of the systems will allow them to be transferred at sea and deployed from vessels of opportunity rather than require dedicated support ships.

Perhaps the most important development is an excavation system designed to operate in a "flying" mode under remote control from the surface. System capabilities will include :-

- location of buried pipelines

- deburial of pipelines up to 36 inch diameter to trench depths of 3 metres in a range of soil strengths up to approximately 100 KPA

- monitor progress of excavation and make final surveys

- backfill trenches

- rapid mobilisation in emergency situations

THE EFFECT OF IMPERFECTION SHAPE ON UPHEAVAL BUCKLING BEHAVIOUR

D.M. Richards

Lloyd's Register, London

ABSTRACT

The upheaval buckling behaviour of heated and pressurised submarine pipelines is reviewed, with particular attention to the role of foundation imperfections.

Exact solutions for continuous foundations are discussed, and the results utilised to produce empirical relationships for analysis and design purposes.

The importance of seabed profile survey to the design process is established, illustrated by results obtained using a recently developed procedure.

INTRODUCTION

Regions of highly localised deformation have recently been discovered in submarine pipelines operating at high temperature.

These disturbances of the regular pipeline profile, involving large plastic strain in the pipe wall, and fracture of coating materials, have been attributed to upheaval buckling (1) (2) (3) (4)

Volume 24: C. P. Ellinas (ed.), Advances in Subsea Pipeline Engineering and Technology, 51–66.

These observations, together with the consequent rather expensive repair procedures, have directed the attention of designers to the need for further research into the mechanics of heated pipeline behaviour.[5]

The concern of the offshore industry has been reflected by the establishment of a joint industry research project to produce design guidelines for upheaval buckling prevention[6] [7].

Meanwhile fundamental research effort has been directed to a detailed investigation of the upheaval process, in order to establish general rules of behaviour which can be applied in the preliminary design situation.

Upheaval buckling is sensitive to both the magnitude and shape of the foundation profile, which in nature is typically of an irregular, random character.

However, any systematic theoretical study must of necessity take as examples idealised shapes of pre-determined analytical form.

The work described here attempts to resolve this dilemma by considering a variety of shapes, for each of which a rigorous analysis of upheaval behaviour through all stages has been carried out.

The envelope of results thus obtained is proposed as a basis for preliminary design, on the understanding that as investigations are extended to further shapes, the envelope may require some extension. Large changes are, however, considered to be unlikely.

In conclusion, the problem of relating these results to the realities of the construction process are considered in a section concerned with seabed survey interpretation.

THE UPHEAVAL PHENOMENON

A pipeline laid upon an uneven foundation, will bear down most heavily upon the high spots or crests of the foundation profile, less so in concave areas. This is due to a combination of negative buoyancy, and the natural tendency of the pipeline to return to its original straight form as constructed.

The introduction of elevated temperature T and internal pressure p will induce an axial compressive driving force given, in fully constrained regions, by

$$P_{eff} = \alpha EAT + p(1-2\nu)D_i^2/4 - T_o \qquad \ldots (1)$$

where
A = pipe steel section area
D_i = internal diameter
E = Youngs modulus
T_o = residual laying tension
ν = Poissons ratio
α = thermal strain coefficient

This force will be reduced in regions where axial slip is possible eg. close to risers.

The effect of this force is to reduce foundation reaction at crests (regions of negative curvature), and to press the pipe down into troughs.

As a consequence there will be changes in the configuration of contact between pipe and foundation.

Generally, these changes will progress smoothly with increasing driving force. However, at a certain load level, no adjacent equilibrium may be possible, and the pipe must jump to a different configuration, usually involving large deformations. This is what is known as an upheaval buckle, the effects of which are usually permanently damaging to submarine pipelines.

PRE-UPHEAVAL

Behaviour before upheaval depends very much on the details of the imperfection shape.

Early studies (1) (2) (4) were concerned with simplified imperfection forms such as the isolated prop, crest and trough, for which closed form solutions were found.

Recently, attention has been directed to somewhat more realistic foundation shapes which, although continuous, are symmetric and assumed to be isolated on an otherwise flat surface (Figure 1). Any given shape of this sort may be scaled by reference to amplitude Δ_f and half wavelength L_o.

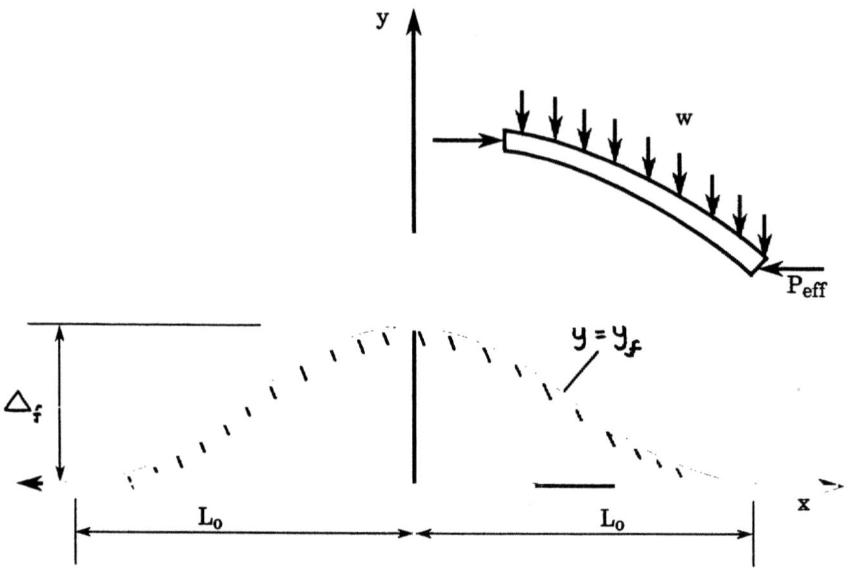

Figure 1: PIPELINE AND FOUNDATION NOTATION

Detailed analysis of these imperfections is complicated by the variety of intermediate equilibrium configurations which must be considered (Figure 2).

However, some general observations to do with upheaval response may be made which relate mainly to the magnitude of imperfection amplitude.

For the isolated prop, which may be regarded as a continuous imperfection of very short wavelength (Figure 2a), behaviour is smooth and continuous until upheaval occurs when contact is lost (Figure 3a).

For small amplitude continuous imperfections (Figure 3b) the upheaval temperature may be reached before detachment is complete.

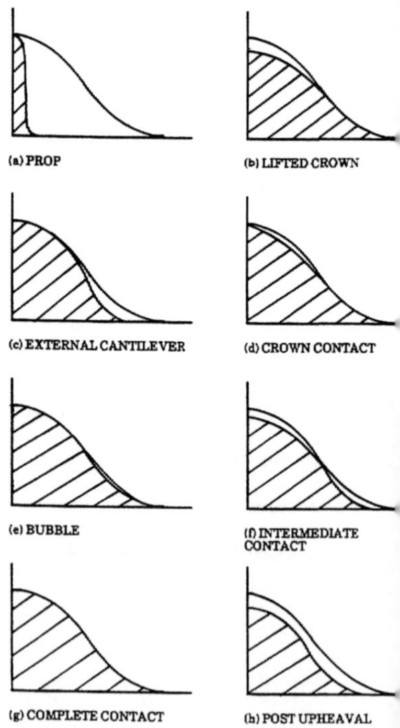

Figure 2: CONTACT CONFIGURATIONS

For somewhat larger amplitudes, complete detachment may again coincide with upheaval (Figure 3c).

For very large amplitudes temperature response may be smooth throughout, with no upheaval as such, although both deformations and stress are nevertheless likely to be unacceptably high.

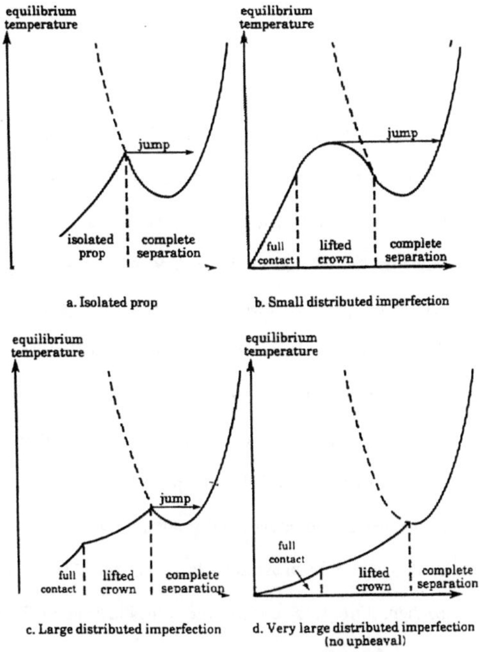

Figure 3: SCHEMATIC EQUILIBRIUM PATHS

POST-UPHEAVAL

The post upheaval configuration (Figure 2h) is common to foundations of all types (see also Figure 3a, b, c, d). In the post upheaval condition, axial force is much reduced and significant axial movements occur as the pipeline feeds into the buckled region.

Up to the upheaval point, however, various studies indicate that changes in axial force and axial friction effects are both negligible.

From an analytical point of view, if one is concerned primarily with behaviour up to upheaval, and the avoidance of upheaval itself, the driving force P_{eff} may therefore be regarded as a constant, separately calculable quantity. This is of great convenience to the designer, who may therefore regard the driving force at the point of upheaval as expressible directly in terms of the required design values of temperature and pressure.

ANALYTICAL SOLUTIONS

A completely analytical solution of the most general class of upheaval problem is impractical.

Thus if one requires to include the effects of non-uniform, non-linear upheaval resistance, irregular foundation shape, foundation flexibility, and plasticity, a finite element routine such as UPBUCK [(6)] is essential.

Some progress has been made, however, by restricting attention to uniformly loaded pipelines resting on rigid foundations, with isolated symmetrical imperfections of analytical form.

These shapes can be taken to correspond to the crests in a real seabed profile.

The essential problem to be confronted by any analytical treatment of this variable contact problem is to match the many different contact configurations to the foundation profile, in order to form a coherent sequence of events.

In order to achieve this for any particular shape, a laborious series of trials involving the numerical solution of simultaneous non-linear relationships is required.

Solutions have been achieved for a number of foundation shapes by constructing consecutive sets of equilibrium configurations selected from the types illustrated in Figure 2.

By matching boundary conditions for each segment either to the foundation or the adjacent segment (see Figure 4) appropriate sets of non-linear equations can be established utilising beam column relationships.

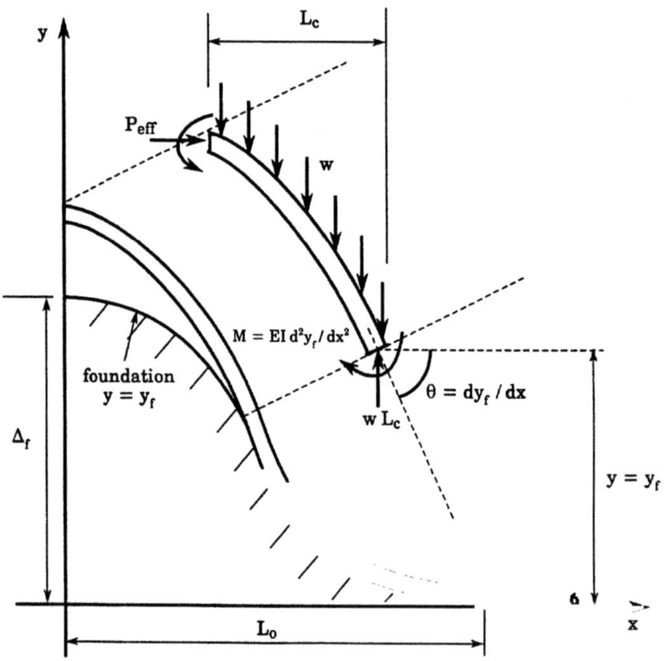

Figure 4: LIFTED CROWN ANALYSIS

The appropriate equilibrium equation is

$$EI.d^4y/dx^4 + Peff.\ d^2y/dx^2 + w = 0 \quad \text{....(2)}$$

This may be arranged in the form

$$H(\gamma_4 + J_o^2\gamma_2) + 1 = 0 \quad \text{....(3)}$$

where $$\gamma_4 = L_o^4/\Delta_f\ d^4y/dx^4\ ;\ \gamma_4 = L_o^2/\Delta_f\ d^2y/dx^2 \quad \text{....(4)}$$

$$H = \Delta_f EI / wL_o^4 \quad \text{....(5)}$$

$$J_o^2 = P_{eff} L_o^2 / EI \quad \text{....(6)}$$

A complete solution of upheaval behaviour for any particular imperfection shape may be found for any given value of the parameter H.

A complete mapping of all solutions for the continuous prop foundation is shown in Figure 5 in the form of trajectories for each H value. This allows progress from the initial unloaded state ($J_0 = 0$) through to upheaval to be traced. In this way, solutions for a number of other shapes have been programmed in the routine LR-SPUR (8), including upheaval, half-sine and pyramid.

Once the complete sequence of equilibrium states has been established, corresponding temperatures may be calculated by comparing total axial strain with the initial (unloaded) configuration.

This process reveals the complete temperature history, including upheaval , as illustrated in Figure 3.

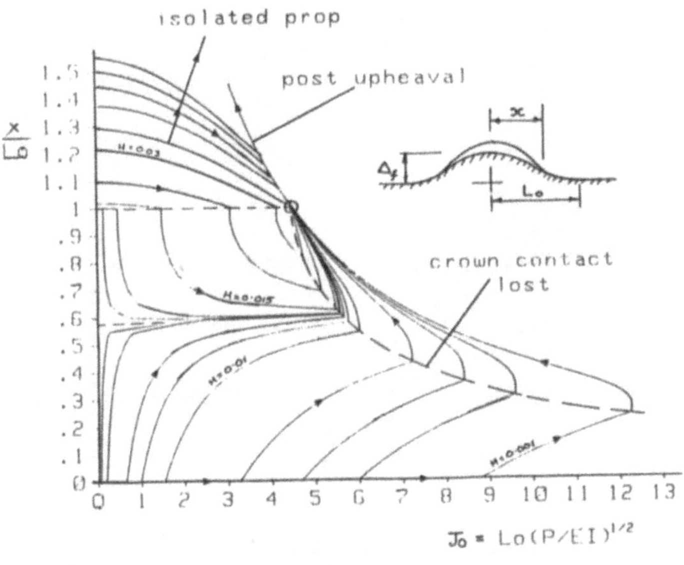

Figure 5: CONTINOUS PROP FOUNDATION SOLUTIONS

UPHEAVAL DESIGN

An important objective of upheaval studies is to establish the level of upheaval resistance required.

In terms of the download required to just maintain contact with the foundation surface, equation (2) may be recast in the form

$$\phi_w = c/\phi_L^4 + d/\phi_L^2 \qquad ....(7)$$

where the down load parameter is

$$\phi_w = wEI/\Delta_f P_{eff}^2 \qquad ....(8)$$

and imperfection wavelength parameter is

$$\phi_L = L_o (P_{eff} / EI)^{1/2} \qquad ....(9)$$

It may be noted from (7) and (8) that download required is proportional to imperfection height.

The relationship with wavelength is more complex.

The constants c and d may be exposed by re-casting (7) in the form

$$\phi_L^2 \phi_w = c / \phi_L^2 + d$$

When numerical results from LR-SPUR are plotted in this form (Figure 6), the nature of equation (7) is confirmed for each chosen foundation shape.

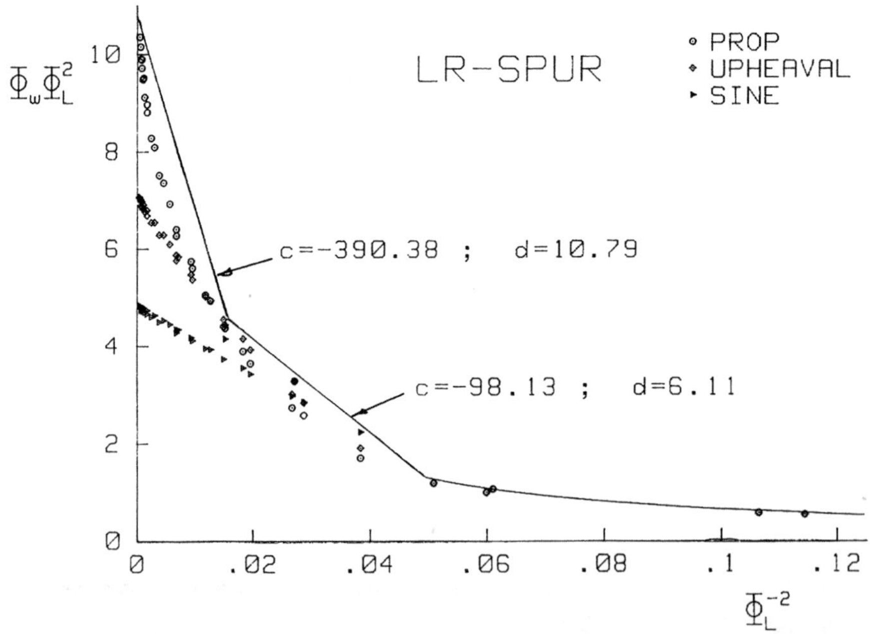

Figure 6: UPHEAVAL BUCKLING NUMERICAL CORRELATION

Many trials with LR-SPUR for a range of practical designs have produced corresponding values of ϕ_w and ϕ_L for different foundation shapes.

These have been plotted in Figure 6 to determine envelope values of the constants c and d, which allows the design curve, shown in Figure 7, to be plotted.

This relationship may be used to estimate download required for a heated pipeline resting on an imperfection of known wavelength and amplitude.

As more shapes are solved, data will be added to increase confidence that these relationships can be assumed to represent the infinite variety of natural occuring seabed formations.

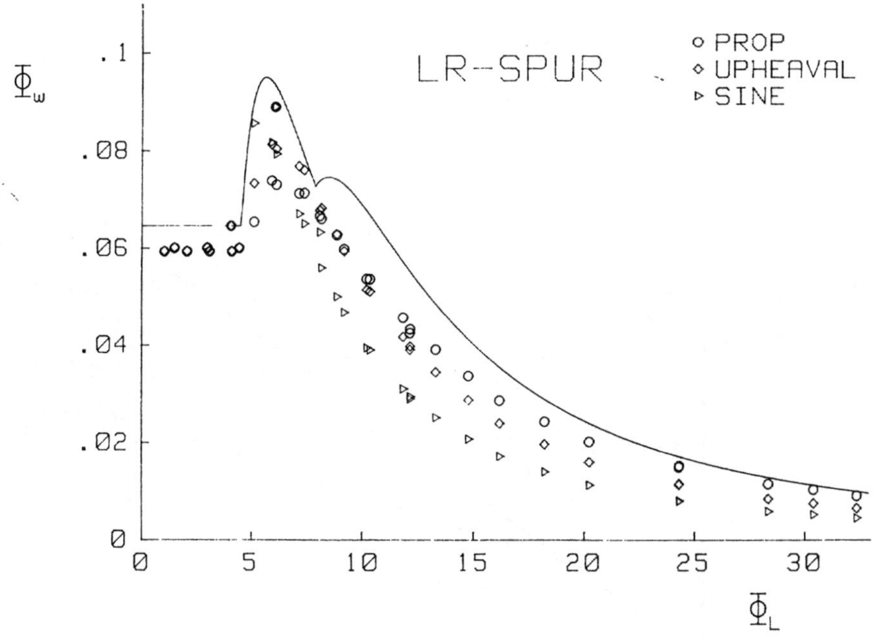

Figure 7: UPHEAVAL DESIGN

UPHEAVAL ANALYSIS

If both foundation shape and available resistance are known, it may become necessary to calculate the safety margin in relation to a given expected driving force, or temperature.

In this case empirical design equation (7) may be recast as an analysis equation. to allow estimation of the allowable driving force. This process leads to

$$\phi_P = 1 / d - c/d.H \qquad ....(10)$$

where

$$\phi_P = P_{eff} \Delta_f / wL_o^{\ 2} \qquad ....(11)$$

and H is given by equation (5).

The empirical values established earlier (Figure 6) give

$$\phi_P = 0.176 + 15.55H \text{ for } 4.4934 < \phi_L < 8.06 \qquad ....(12)$$

$$\phi_P = 1.104 + 35.73H \text{ for } \phi_L < 8.06 \qquad ....(13)$$

For $\phi_L < 4.4934$, we have isolated prop behaviour, for which

$$P_{eff} = 3.962 (wEI / \Delta_f)^{1/2} \qquad ....(14)$$

The above results are shown plotted in Figure 8.

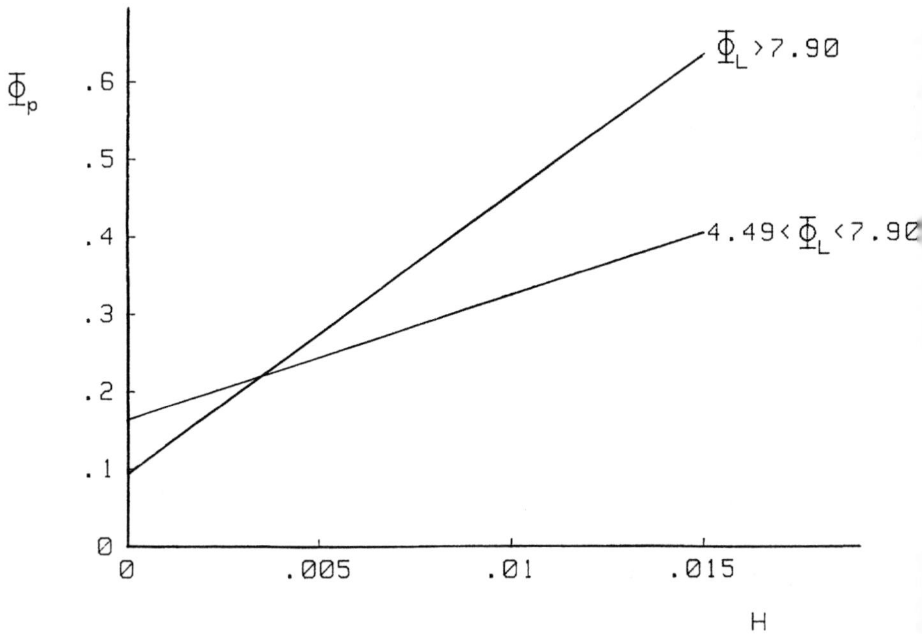

Figure 8: UPHEAVAL ANALYSIS

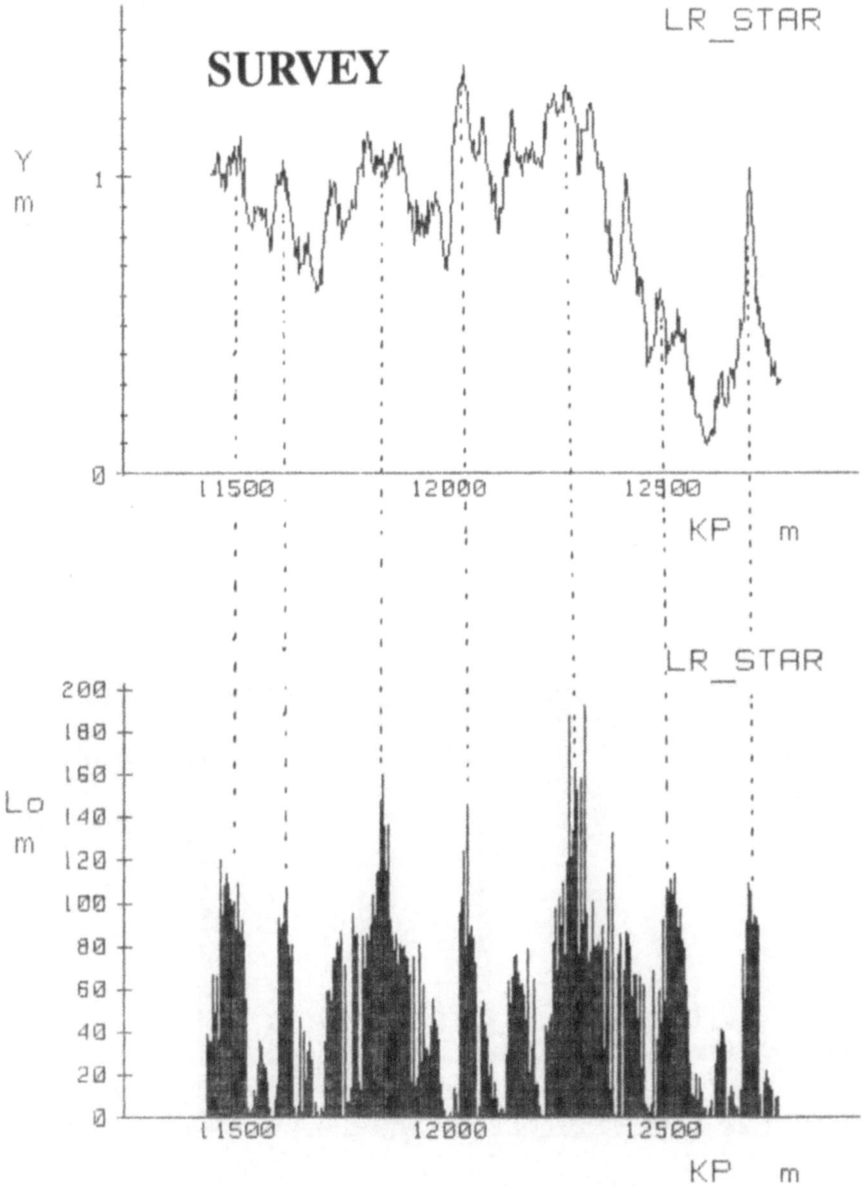

Figure 9: SURVEY INTERPRETATION: SEABED FEATURE WAVELENGTHS

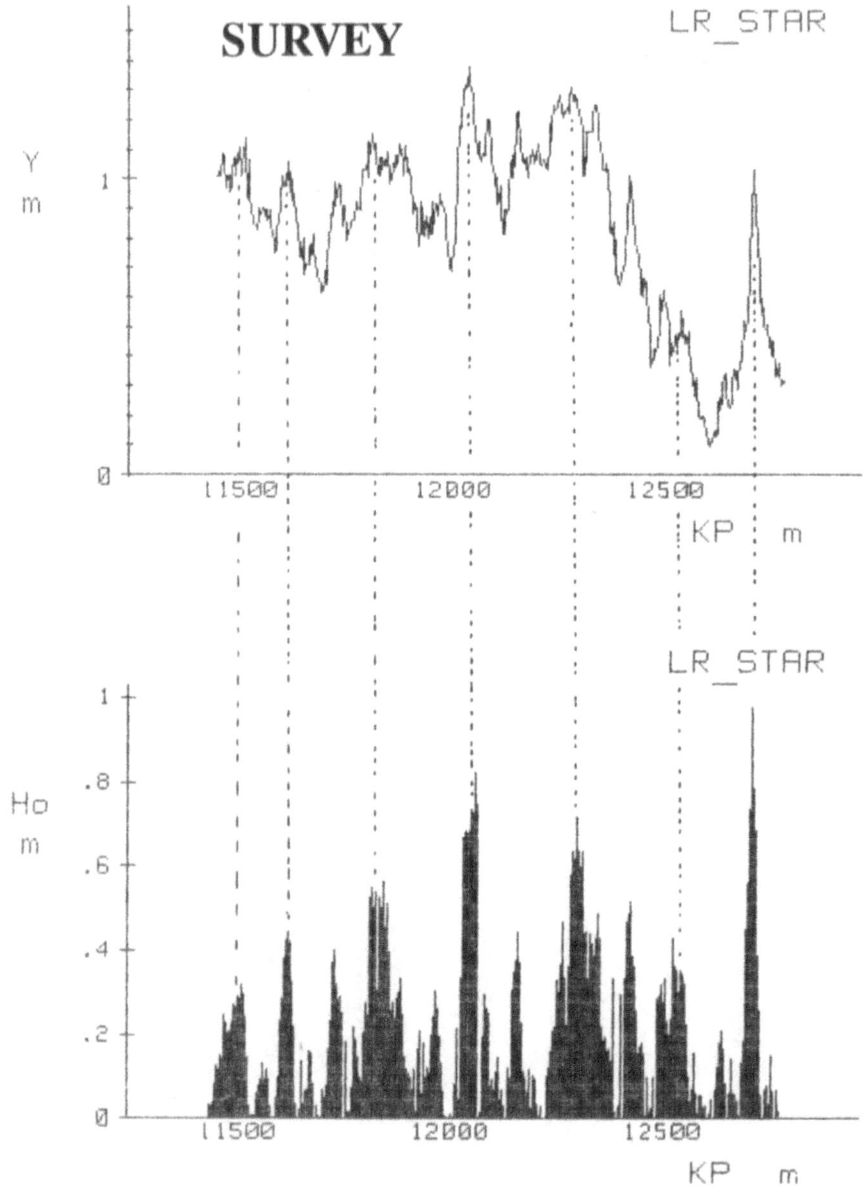

Figure 10: SURVEY INTERPRETATION: SEABED FEATURE AMPLITUDE

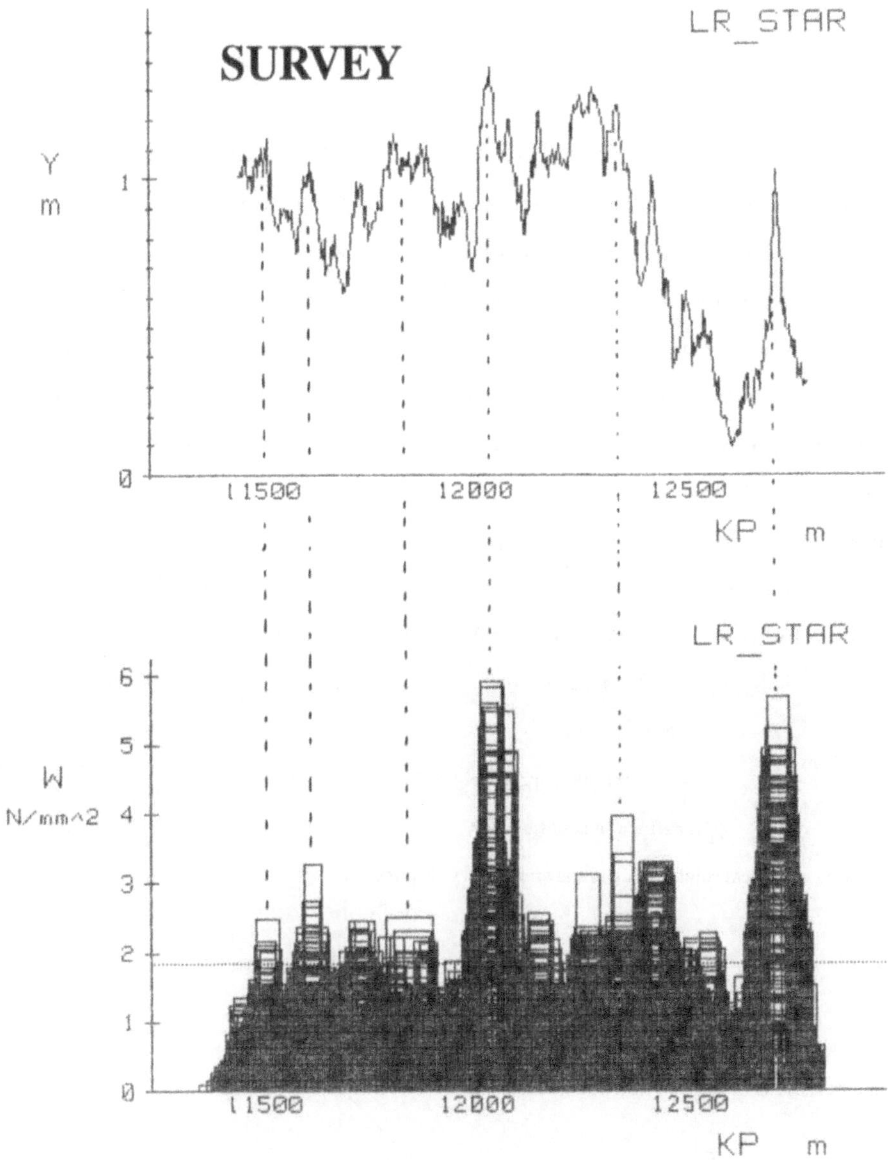

Figure 11: SURVEY INTERPRETATION: DOWN LOAD REQUIRED

SEABED SURVEY

An important consideration in practical upheaval design is the determination of the amplitude, wavelength, shape and location of imperfections actually existing beneath the pipeline as constructed. The evidence available will be in the form of a survey consisting of a series of vertical ordinates relative to some datum, more or less equally spaced along the pipeline route.

The interpretation of this data is a major task, requiring a routine such as LR-STAR [5][9] This program, developed explicity for the purpose, identifies and locates seabed features on the basis of survey data, giving information on imperfection wavelength and location, together with preliminary estimates of the distribution of soil cover required to prevent upheaval.

Typical results are shown plotted in Figures 9,10, and 11.

The review of results is aided by rearranging equation (7) to give an indication of the imperfection amplitude which can be tolerated without additional soil cover. This is given by

$$\phi_\Delta = \phi_L^4 / (c + d\phi_L^2) \qquad \text{....(15)}$$

where $\phi_\Delta = \Delta_f P_{eff}^2 / w_o EI$

w_o = pipeline submerged weight

ϕ_L is defined in equation (9)

The resulting empirical relationship is shown plotted in Figure 12.

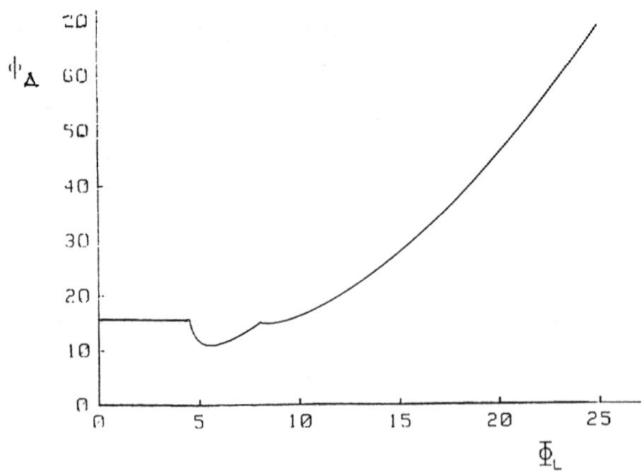

Figure 12: AMPLITUDE LIMIT FOR PIPE WEIGHT ALONE

CONCLUSIONS

Aspects of upheaval buckling behaviour have been reviewed in relation to analysis and design procedures.

Empirical relationships suitable for preliminary design have been developed, based on certain exact solutions of upheaval behaviour involving continuous foundation imperfections.

The importance of a detailed route survey and appropriate interpretation procedures for the design process, has been established.

ACKNOWLEDGEMENT

The author would like to acknowledge the contribution made by his colleagues L. Freeman and R. Pearson in the preparation of this paper.

REFERENCES

1. Kerr, A.D.
'On the stability of the Railroad Track in the Vertical Plane'
Rail International V5 pp 131 - 142, 1974

2. Courbon, J.
'The Equilibrium of a Heated Beam Resting over a Rigid Horizontal Plane'
Journal de Mecanique Appliqueee V4 N4 1980

3. Boer, S., Hulsbergen,C.H., Richards, D.M., Klock, A., and Biaggi, J.P.

'Buckling Considerations in the Design of the Gravel Cover for a High Temperature Oil Line'

OTC 5294 Houston 1986.

4. Richards, D.M., and Andronicou, A.

'Seabed Irregularity Effects on the Buckling of Heated Submarine Pipelines'

Offshore Holland Amsterdam Nov. 1986.

5. Lloyds List International 31 January 1990, p4.

6. Guijt, J.

'Research on Upheaval Buckling of Offshore Pipelines'.

O.P.T. Paris 1990

7. Palmer, A.C., Ellinas, C., Richards, D.M.

'Design of Submarine Pipelines Against Upheaval Buckling'

OTC 6335 Houston 1990

8. LR-SPUR: Submarine Pipeline Upheaval Routine

Lloyd's Register, London 1989.

9. LR-STAR: Seabed Topography Analysis Routine

Lloyd's Register, London 1989.

BEHAVIOUR OF ANCHOR REINFORCEMENT IN OFFSHORE PIPELINES

A. Bouazza and T. W. Finlay

Civil Engineering Dept., Glasgow University, Glasgow (Scotland).

INTRODUCTION

Plate anchors are an economical alternative to gravity and other embedment anchors for resisting uplift forces in the marine environment. Generally the plate is welded to a steel stem with an eye hook which accepts the external load. The ease with which plate anchors can be installed (Mc Cormick, 1979), and the fact that grouting is not required make them ideal for use underwater, to secure pipelines, moorings, and cables to the seabed. Bobbit And Clemence (1987) reported the use of anchor sets having capacities of 45 KN to 150 KN to stabilize large diameter pipelines in Indonesia as illustrated in fig. 1a. Similarly on land, geogrids have been used to enhance the uplift capacity of buried pipelines (Selvadurai, 1989). To date no field cases have been reported from the U.K. sector. Field experiments are expensive to set up, but it is the authors' hope that when a reliable theory is produced on the basis of model tests, such those described in this paper some full-scale tests can be carried out to verify the theory.

Volume 24: C. P. Ellinas (ed.), Advances in Subsea Pipeline Engineering and Technology, 67–76.

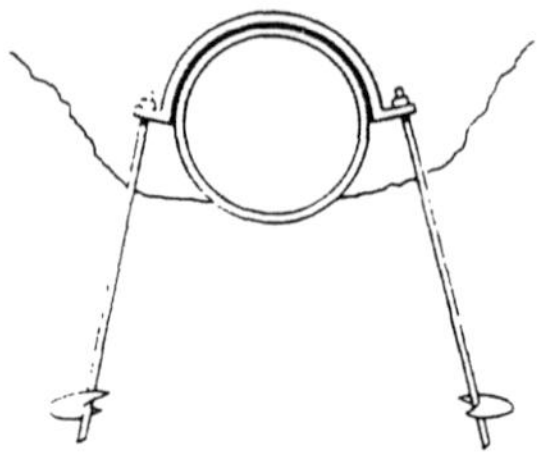

Fig 1a Pipeline anchor (after Bobbit & Clemence, 1987).

Previous research into the uplift capacity of plate anchors embedded in cohesionless soil has shown that two distinct categories of anchor can be identified; shallow anchors and deep anchors. In the former, the anchor is installed close to the surface of the soil, and the failure surface in the soil extends from the tip of the anchor to the ground surface with significant surface movements. An increase in depth of embedment results in another type of failure in which the failure surface does not extend to the surface but instead forms locally around the anchor. This type of failure mechanism exemplifies the deep anchor mode.

This paper presents the results of some model tests on shallow horizontal plate anchors and shows the effect on uplift resistance of varying the embedment depth and the relative density of the sand in which the anchor is embedded. The tests described are part of a wider study which has been in progress for some time at Glasgow University. The complete study is aimed at assessing the behaviour of simple circular plate anchors embedded in sands and clays, and subjected to static and cyclic wave loading. The following references describe some of this work: Ponniah (1984) and Stewart (1988).

The aim of the wider study is to determine the parameters which affect pull-out resistance of a simple anchor, and eventually to produce a theory which will lead to more accurate predictions than are currently available. The comparisons of the results from the tests described in this paper with current theories ullustrate the variations in predicted uplift which can occur.

BACKGROUND

Most of the approaches used to predict anchor holding capacity are based upon the limit equilibrium concept. The major difference amongst the various methods involves the assumed shape of the slip surface developed within the failing soil mass. Limit equilibrium conditions compute the ultimate anchor uplift capacity (P) as the combined effects of the weight of soil contained within the assumed failure surface, the shearing resistance of the soil integrated over the failure surface, and the dead weight of the anchor structure itself.

A review of most of the theoretical studies used for evaluation of the ultimate uplift capacity of plate anchors embedded in sand has been given by Bouazza (1990). It is not the intention of this paper to review all pertinent theories, however, but those for plate anchors provided by Meyerhof and Adams (1968), Vesic (1971), and Rowe and Davis (1982) will be briefly discussed below since these are the most widely referred to in literature. The methods presented are used in practice in the same way as Terzaghi's (1948) bearing capacity formula, by the use of dimensionless parameters.

Terms used in anchor research are the depth/diameter ratio (D/B) and the breakout factor (Nu); the former is defined as the depth (D) from ground surface to the tip of the anchor, divided by the plate diameter B. The latter is defined as the ultimate pressure on the anchor plate (Pu/A), where A is the anchor area, divided by the effective overburden pressure (γ D) where γ is the soil unit weight. Thus, for cohesionless soils:

$$Nu = (Pu/A)/\gamma D \quad .... \quad (1)$$

The use of a non-dimensional breakout factor allows the reader to compare different theories as well as test results from anchor tests with dissimilar anchor sizes, soil densities, and embedment depths.

<u>Meyerhof & Adams theory (1968)</u>

The above authors proposed a general uplift capacity theory which was complementary to the general bearing capacity theory proposed by Meyerhof. A curved failure surface was adopted extending vertically above the perimeter of the anchor and solved for a strip footing. This theory was then applied to circular and rectangular anchors by the introduction of a shape factor term (s) into the general expression. The ultimate uplift capacity in general shear was computed by summing the shearing resistance from cohesion and passive earth pressure at an angle from a vertical cylindrical surface through the anchor edge and is determined as:

$$P = (\pi/2)\ B\ \gamma\ D^2\ s\ K\ \tan\varphi \ \ldots\ldots\ldots\ldots \ (2)$$

Hence the breakout factor may be written as:

$$Nu = 2(D/B)\ s\ K\mu\ \tan\varphi + 1 \ \ldots\ldots\ldots\ldots \ (3)$$

where, $K\mu$ is the coefficient of lateral pressure and φ is the angle of internal friction.

Vesic theory (1971)

Vesic analysed the problem by using solutions of spherical and cylindrical cavities near the surface of a semi infinite rigid plastic solid. The ultimate pressure, Pu, needed to breakout a spherical cavity close to the surface was given in terms of a spherical cavity expansion factor Fq.

$$Pu = \gamma\ D\ [Fq + 1/3\ (B/D)] \ \ldots\ldots\ldots\ldots \ (4)$$

The breakout factor can then be written as:

$$Nu = Fq + B/3D \ \ldots\ldots\ldots\ldots\ldots\ldots\ldots \ (5)$$

where Fq is function of the shape, relative depth and angle of internal friction.

Rowe & Davis theory (1982)

The above authors described a theoretical investigation on cohesionless soils in which consideration was given to the effect of anchor embedment, friction angle, dilatancy, initial stress state K_0 and anchor roughness. Their theoretical approach was based on an elastoplastic finite element analysis. The soil was assumed to have a Mohr Coulomb failure criterion and either an associated or a non associated flow rule. The anchor breakout factor was given as:

$$Nu= F_{\gamma}\ R_{\psi}\ R_{k}\ R_{r} \quad \ldots\ldots\ldots\ldots\ldots (6)$$

Where F_{γ} is a basic anchor capacity factor, and R_{ψ} R_k R_r are correction factors for the effect of soil dilatancy, initial stress state and anchor roughness respectively.

TESTING PROGRAMME

Model anchor pull out tests were performed in Leighton Buzzard sand which exhibited the following engineering properties:

Mean grain size D_{50} (mm): 0.8

Coefficient of uniformity: 1.8

Specific Gravity: 2.65

Dominant mineral: Quartz

Shape: Subrounded

Max porosity (%): 44.2

Min porosity (%): 33.2

The tests were performed in a steel tank 500 mm diameter and 700 mm deep, large enough for tests to be carried out without any side effects. The model anchor was made from brass, 37.5 mm diameter and 3 mm thick, with an anchor shaft 6mm diameter screwed into the brass disc to make the anchor unit. At the start of each test a 75 mm thick layer of sand was compacted into the bottom of the tank. The anchor plate was then placed in position on top of this layer, and the sand rained in layers into the tank from a required height to produce a constant density (Kolbusewski & Jones, 1961). The process of sand laying was repeated until the anchor plate was buried to the required depth. Details of the soil densities and the corresponding relative densities and friction angles are given in table 1.

The anchor unit was attached to a load cell which was connected to a pneumatic

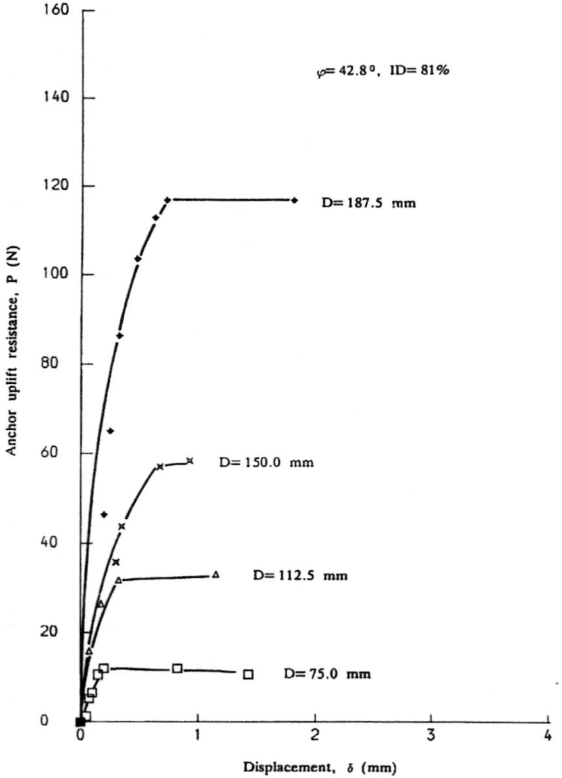

Fig. 1b Effect of anchor depth on load/displacement relationships for a circular anchor embedded in dense sand.

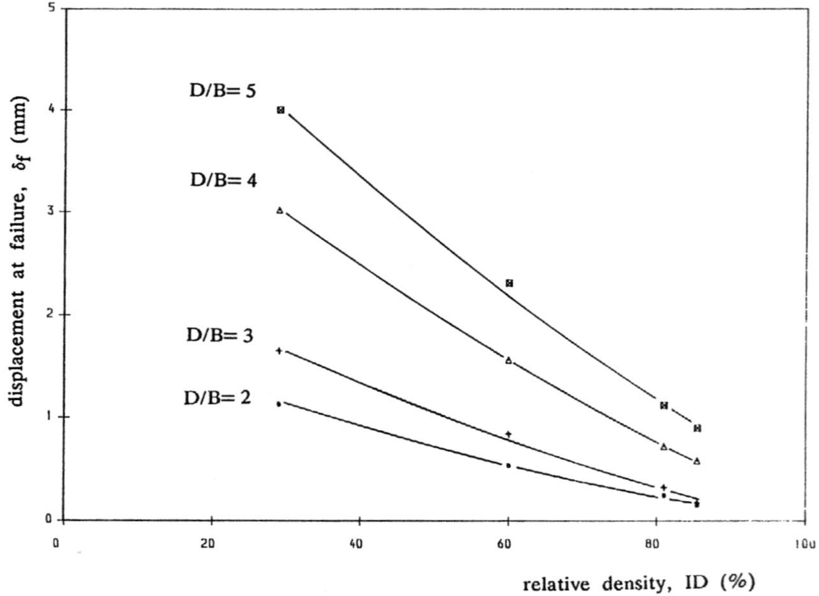

Fig. 2 Anchor displacement at failure– relative density relationship.

loading device to apply a static loading, under load control (incremental loading). As the applied load approached the failure value, the load increment was reduced to give a more clearly defined load value. The ultimate load was taken as that measured at the last increment before total failure occurred (Fadl, 1981). A linear variable displacement transducer recorded the displacement of the anchor throughout the tests and this information, plus the load cell reading, was logged at intervals throughout each test.

Table 1 Sand characteristics in the present study

sand	density (KN/m^2)	Relative density (%)	φ (^{0})
	17.28	85.5	43.7
Leighton	17.15	81.0	42.8
Buzzard	16.57	60.0	39.0
	15.63	29.0	33.8

TEST RESULTS AND DISCUSSION

Typical load displacement curves at various embedment ratios are shown in fig. 1. It is seen that the ultimate uplift capacity increases with increase in depth of embedment. Examination of this figure also shows that the anchor exhibited a stiff, near linear, response for the early part of the loading range and, in the vicinity of the ultimate load, the displacement increased significantly before the peak load was attained. The load displacement diagram also demonstrates that the displacement of the plate anchor at which the ultimate static resistance occurs is small and ranges from 0.2% to 2.4% of the initial depth. This indicates that only a small displacement of the anchor is required to mobilize the frictional resistance of the sand. Fig. 2 shows the effect of sand densification on the vertical movement of the anchor. It was found that the movement at failure of the anchor surrounded by loose soil was approximately 1.5 times greater than the same anchor embedded in dense soil. This can be explained by the fact that when the sand is denser, the degree of interlocking is high therefore the load from the anchor transferred to the sand was only required to break out the soil mass rather than to first densify it. The variation of the breakout factor with depth/embedment is shown in fig. 3 for relative densities varying from 29% to 85.5%. A clear pattern emerges from this figure where it can be seen that the effect of the degree of densification of the soil and D/B is to greatly increase the uplift resistance of the anchor. Depending on the depth of embedment, increasing the relative density of the sand from a loose (ID= 29%) to a very dense condition (ID= 85.5%) increases the breakout factor by 2 to 3 times.

Analysis of Prediction Methods

In order to compare the present experimental results with various existing theories, the experimental breakout factors are shown in fig. 3 along with the theoretical plots as obtained from the theories of Meyerhof & Adams, Vesic and Rowe & Davis.

For the dense soil (φ= 43°, ID= 85.5%) the method described by Meyerhof & Adams seems to correlate best with the test data. However, in the loose state the theoretical values were found to be higher than the experimental values. The other methods employed to predict the sand tests are seen to underpredict the failure load by as much as 300%, thus indicating the insensitivity of these methods to the variations of φ and ID.

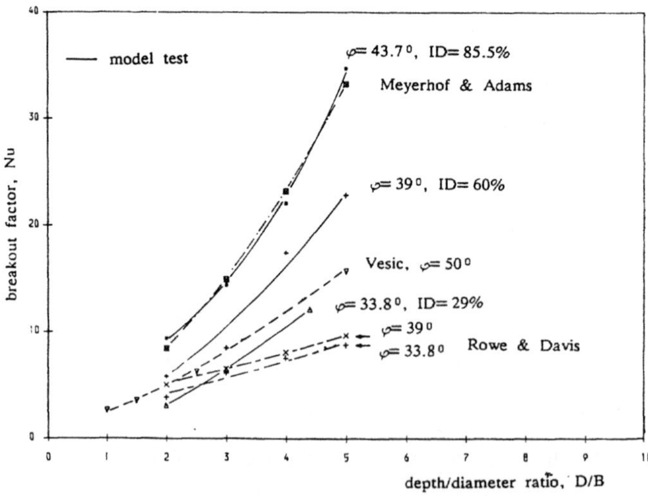

Fig. 3 Breakout factor vs. depth/diamter ratio: comparing predicted behaviour with model test results in sand.

CONCLUSIONS

Results of laboratory model tests for ultimate uplift capacity of shallow circular anchors embedded in loose, medium and dense sand have been presented. Based on the present study, the following conclusions can be drawn:

1. Load displacement behaviour of anchor plates and the breakout factor depend on the depth of embedment and relative density.
2. The theory developed by Meyerhof & Adams fits the available test data in dense sand better than other methods considered. Vesic and Rowe & Davis theories severely underestimate the breakout factor.

ACKNOWLEDGEMENT

The research described herein was cosponsored by CNERIB (Algiers, Algeria) and the British Council, this support is gratefully acknowledged.

REFERENCES

Bobbit, D. E. and Clemence, S. P. (1987). Helical anchors:application and design criteria. Proc. 9th Southeast Asian Geotech. Conf., Bangkok (Thailand), Vol. 1, pp. 105–120.

Bouazza, A. (1989). Effect of sand properties on the uplift resistance of plate anchors. PhD Thesis (forthcoming). Glasgow University, Glasgow (Scotland).

Fadl, M.O. (1981). The behaviour of plate anchors in sand. PhD Thesis. Glasgow University, Glasgow (Scotland).

Kolbuszewski, J.J. and Jones, R.H (1961). The preparation of sand samples for laboratory testing. Proc. Midland Soil Mech. Fdn. Engng. Soc., Vol. 4, pp. 107–123.

Mc Cormick, M.E. (1979). Anchoring systems. Eds. Pergamon press.

Meyerhof, G.G and Adams, J.D. (1968). The ultimate uplift capacity of foundations. Canadian Geotechnical Journal. Vol. 5 (4), pp. 225–244.

Ponniah, D.A. (1984). Behaviour of plate anchors in cohesive soils under static & cyclic loads. Ph.D Thesis. Glasgow University, Glasgow (Scotland).

Rowe, R.K. and Davis, E.H. (1982). The behaviour of anchor plates in sand. Geotechnique 32. No 1, pp. 25–41.

Selvadurai, A.P.S (1989). Enhancement of the uplift capacity of buried pipelines by the use of geogrids. Geotechnical Testing Journal, Vol.12, No 3, pp. 211–216.

Stewart, W.M. (1988). Plate anchors in sand under static & cyclic loads. Ph.D Thesis. Glasgow University, Glasgow (Scotland).

Terzaghi, K. (1948). Theoretical Soil Mechanics. John Wiley & Sons, New York (USA).

Vesic, A.S. (1971). Breakout resistance of objects embedded in ocean bottom. Proc. ASCE, Jour. Soil Mech. Fdn. Div., Vol. 97, No SM 9, pp. 1183–1205.

Part II

Flexible Pipelines

INCREASED RELIABILITY THROUGH A UNIFIED ANALYSIS TOOL FOR BONDED AND NON-BONDED PIPES.

S. A. Løtveit and Ola Often

BrasNor A.S. , p.o. box 24, N-1364 Hvalstad, Norway

1 ABSTRACT

Methods to predict stresses in flexible pipe under both combined loading and damaging loads on flexible pipes exist, these analytical tools are generally proprietary to the manufacturers and usually incorporate empirical factors linked to test results from specific pipe types. This poses a major problem in relation to incorporating the use of these tools into general pipe design guidelines.

All flexible pipes are essentially layered structures incorporating elastomer or thermoplastic layers to contain pressure and using spiralled armor to take loads. The ability of the reinforcing spirals to relocate under bending gives the pipe its bending flexibility. Two main types of pipe exist: bonded pipes, where all armoring is embedded in and bonded to an elastomer compound, and non-bonded, where the helices can slide

Volume 24: C. P. Ellinas (ed.), Advances in Subsea Pipeline Engineering and Technology, 79–110.

against each other and adjoining layers.

The difference in behavior of bonded and non-bonded flexible pipe is discussed. A general theoretical formulation of the stiffness relations for layered flexible pipe is presented, and it is shown how to model a pipe in a way which is consistent for both bonded and non-bonded pipes.

This theory has been implemented in the computer program FLEXPIPE which has been verified against test results and calculations made by the manufacturers. Some comparisons with test results are presented which indicate that using a non-bonded load transfer model for a bonded case does not produce the required level of accuracy, and that the bonded model as implemented in FLEXPIPE is required.

The aim should be to bring flexible pipes to a documented safety level equivalent to other parts of the overall production system. Present safety factors are typically based on experience, industry practice and testing. A revised design approach should be adopted which takes into account combined loading effects, accuracy of the analytical prediction and spread in test data. A such design approach requires use of a well documented and available computer program. A case study is included to show the advantages of using a consistent design approach for optimization of a deep water pipe, layer by layer.

2 DESCRIPTION OF THE FLEXIBLE PIPE STRUCTURE

The term flexible pipe can cover a lot of different products, from garden hose to corrugated steel pipe. The purpose of this theory is however to describe the behavior of a pipe which provides leak-proofness, flexibility and strength through a combination of polymer elastic material and helically wound reinforcement. The spiral reinforcement takes both hoop and axial tension forces, the angle and location of reinforcement layers determining the distribution of forces.

A common special case is that in which all reinforcement is laid with the same lay angle (although cross-laid). As the ratio between hoop force and axial tension is constant under pure internal or external pressure, there is an angle (54.7 degrees) at which the forces in the pipe are balanced under such loading, avoiding excessive elongation or radial expansion. This angle is called the neutral angle.

Due to both its manufacturing process and function, the flexible pipe is a layered structure, incorporating alternating layers of polymer material and reinforcement. The polymer material may be thermoplastic (polyamide, polyethylene) or elastomer (nitrile rubber, chloroprene rubber etc.). Reinforcement elements may be steel helices (independent or interlocked), steel cord, aramide fiber ropes, fabric of various materials etc. Special layers such as gas tight corrugated steel liner is not considered here.

Two main principles exist for flexible pipe construction:

Non-bonded pipe.

The layers interact only through contact pressure (or gaps) and some friction. The void between reinforcement helices is air-filled. The sealing (thermoplastic) layers have sufficient strength to distribute pressure to spirals in reinforcement layers.

Bonded pipe.

Each reinforcement layer is embedded in and bonded to surrounding rubber. Adhesion between the layers will restrain gaps between layers and also prevent slip during bending. The rubber between spirals will also transfer forces normal to each spiral. The low stiffness of rubber gives the rubber a tendency to act more like a fluid than a stiff membrane. This is partly counteracted by use of fabric reinforcement.

Under axi-symmetrical loading there is no or insignificant slip between layers, even in an non-bonded pipe. Under bending, however, the non-bonded pipe will see large slip, or relative tangential/axial movement between different helix layers. This relocation allows the pipe to bend without imposing significant stress in the helices. For the bonded pipe, a similar relocation will occur, but this will be counteracted by shear forces in the elastomer between the layers.

3 GENERAL STIFFNESS FORMULATION FOR AXI-SYMMETRICAL LOADS

3.1 Introduction

In order to describe the specific theory for the bonded flexible pipe, a short description of the general formulation of the flexible pipe structural behavior is necessary. The following theory is meant to describe the force and deformation relations in a flexible pipe under axi-symmetrical loads, i.e. pressure, axial tension and torsion moment, individually or in combination.

The general theory described below is an alternative formulation of a theory for the structural behavior of flexible pipes, non-bonded flexible pipes in particular, which has been published by various authors, ref. /1/,/2/,/3/ and /4/.

3.2 Analysis Methodology

The pipe is described as a set of independent layers, each layer being a thin-walled cylinder described geometrically by a mean diameter D and a thickness t (see fig. 1). Length is not defined as the pipe is assumed uniform along its length (not necessarily true near the end fittings). The assumption of thin layers is a convenient simplification which gives insignificant errors for all realistic examples.

The layer may be subject to internal pressure P_i , external pressure P_e, axial force F_a and torsion moment M_t, see fig 1. The surface pressures may be fluid pressure and/or contact pressure from adjoining layers (this may be negative for a bonded pipe). The constraint of the end-fitting and the assumption of uniformity along the length leads to equal axial elongation and rotation angle for all layers. The layers interact through surface contact pressure and by transfer of axial tension and torsion moments through the end-fittings.

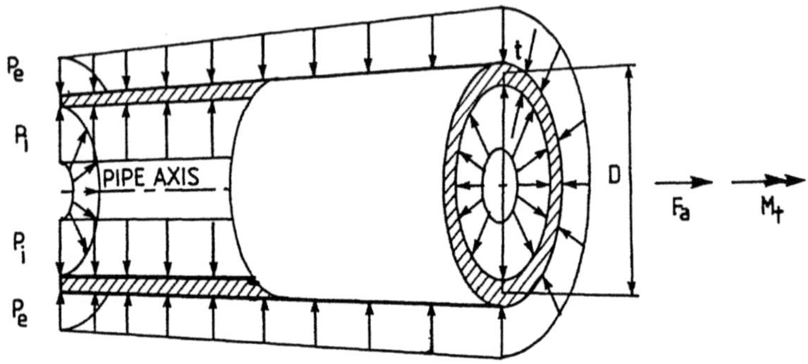

Figure 1 Layer Element

The approach adopted is that of the common finite element method, describing a linear relationship between forces and deflections on element level, and using kinematic compatibility and force equilibrium between the elements to assemble a set of linear equations between forces and deflections which describe the structural behavior of the whole system. This set of equations may then be solved by common matrix operations. This approach has several advantages, some of which are:

* By following common FEM practice, the energy principles are inherently adhered to.

* The theory is familiar and easily accessible to engineers with structural background.

* Matrix formulations makes both theoretical work and computer implementation easier.

3.3 Layer Element Stiffness Relations

3.3.1 Definition of Strains

The deformation of a layer may be defined by four independent strains:

Axial strain:

$$\varepsilon_a = \delta L/L$$

Tangential (hoop) strain (as circumference is proportional to diameter, this is equal to relative increase in diameter):

$$\varepsilon_t = \delta\pi D/\pi D = \delta D/D = \delta R/R$$

Thickness (radial) strain:

$$\varepsilon_r = \delta t/t$$

Torsional shear strain:

$$\gamma = \delta\phi/L_o \cdot D/2$$

where (see fig. 2):

L_o = length in axial direction

δL = elongation over length L

D = mean diameter

δD = increase in diameter

R = mean radius

δR = radial displacement

t = wall thickness

δt = increase in wall thickness

$\delta\phi$ = pipe rotation angle (in radians) over pipe length L_o

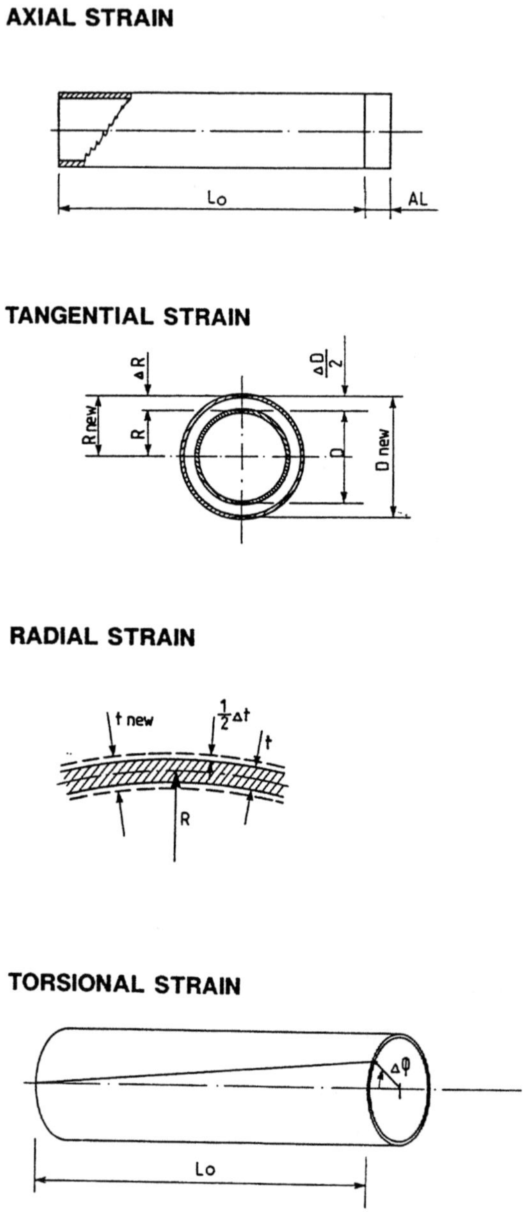

Figure 2 Definition of strains

3.3.2 Definition of Stresses

The mean stresses in the layer may be defined as:

$$\boxed{\sigma_a = F_a/\pi Dt}$$

$$2t\sigma_t = P_i(D-t) - P_e(D+t)$$

$$= (P_i - P_e)D - (P_i + P_e)t$$

$$\sigma_t = (P_i - P_e)D/2t - (P_i + P_e)/2$$

$$\boxed{\sigma_t = \delta P \cdot D/2t - P}$$

$$\sigma_r = -(P_i + P_e)/2$$

$$\boxed{\sigma_r = -P}$$

$$\tau = M_t/\pi Dt \cdot 1/R$$

$$\boxed{\tau = 2M_t/\pi D^2 t}$$

where (see fig. 3):

σ_a = axial stress

F_a = axial force

D = mean diameter

t = wall thickness

P_i = internal pressure

P_e = external pressure

δP = pressure differential (positive outwards from pipe axis)

P = mean pressure over thickness

M_t = torsion moment

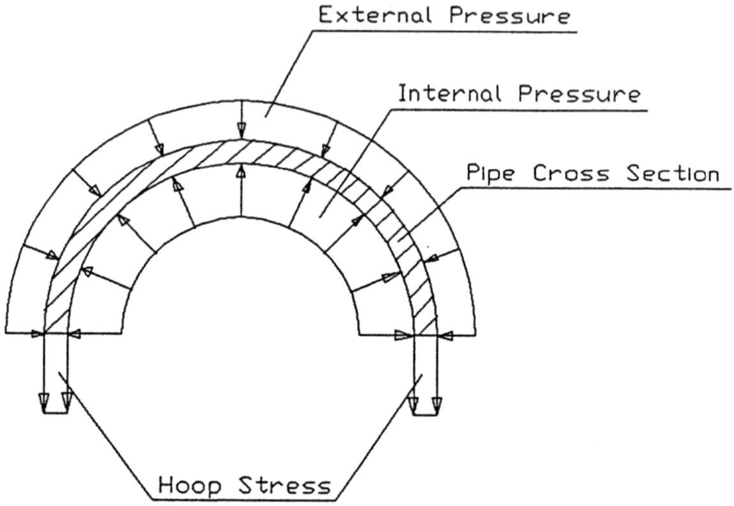

Figure 3 Pressures and Hoop Stress

3.3.3 General Layer Stiffness Formulation

It may be assumed that for any type of layer, it is possible to write a relation between section forces and pipe strains as defined in sections 3.3.1 and 3.3.2:

$$\begin{bmatrix} \sigma_a \\ \sigma_t \\ \sigma_r \\ \tau \end{bmatrix} = \mathbf{k} \cdot \begin{bmatrix} \varepsilon_a \\ \varepsilon_t \\ \varepsilon_r \\ \gamma \end{bmatrix}$$

where **k** is the force-strain stiffness matrix, σ is the mean section stress vector and ε is the mean strain vector, all defined relative to global pipe axis. Using the equations between mean stresses and section forces from 3.3.2, we may transform the stiffness matrix to section forces:

$$\underbrace{\begin{bmatrix} F_a \\ \delta P \\ P \\ M_t \end{bmatrix}}_{f} = \begin{bmatrix} \pi Dt & 0 & 0 & 0 \\ 0 & 2t/D & -2t/D & 0 \\ 0 & 0 & -1 & 0 \\ 0 & 0 & 0 & \pi D^2 t/2 \end{bmatrix} \cdot k \cdot \begin{bmatrix} \varepsilon_a \\ \varepsilon_t \\ \varepsilon_r \\ \gamma \end{bmatrix}$$

By using a simple set of relations between the tangential strains at the surfaces and mean tangential strain and thickness strain, and likewise between surface pressures and pressure differential and mean pressure, the layer stiffness relation may be transformed to the following form:

$$\underbrace{\begin{bmatrix} F_a \\ P_i \\ P_e \\ M_t \end{bmatrix}}_{f_g} = k_g \cdot \underbrace{\begin{bmatrix} \varepsilon_a \\ \varepsilon_{ti} \\ \varepsilon_{te} \\ \gamma \end{bmatrix}}_{\varepsilon_g}$$

where:

f_g = Layer global force vector.

k_g = Layer global stiffness matrix.

ε_g = Layer global strains.

P_i = Pressure on internal surface.

P_e = Pressure on external surface.

ε_{ti}= Tangential strain on internal surface.

ε_{te}= Tangential strain on external surface.

Note again that tangential strain is equal to relative increase in radius.

When assembling the system stiffness matrix and force vector for all layers, this form is more convenient to work with. The surfaces can be treated equivalent to nodes in a normal frame element analysis. Pressure is here defined with positive outwards direction, even on outside surface.

3.4 Solving The System Of Equations

Given the layer stiffness relations for all the layers, the total system of equations may be assembled following common finite element method practices, satisfying kinematic constraints (surface contact, end fitting constraints) and force equilibrium. For a non-bonded pipe, the actual number of degrees of freedom will depend on the gap/contact conditions, which depends on the loading. Therefore, the solution algorithm for this type of pipe must be iterative, checking for negative surface pressure or surface overlap and updating gap constraints until equilibrium is achieved.

Fluid pressure must be applied at the degree of freedom associated with the actual pressure barrier. As there is only one axial and one torsional degree of freedom, the applied axial tension and torsion moment is always applied globally. In the case of pressure loads (external or internal) the pressure-induced axial force must be added to the axial tension.

4 LOAD TRANSFER PRINCIPLES

4.1 General

In the case of the simple tube with isotropic material, the mean stresses as defined above are also actual stresses, and the layer stiffness relations may be derived by a simple application of Hooke's law for tri-axial stresses. In the case of helix armoring layers however, the relation between mean stresses and strains must be derived from the stresses and strains in both the helix and surrounding (air or rubber) materials. The load transfer mechanism is also dependent not only on the geometry and materials of the layer itself, but the nature of adjoining layers as well.

4.2 <u>Non-bonded Helix Layer</u>

The layer consists of a number of parallel, helically wound, equally spaced strips or wires with air between them. The layer does not contain fluid pressure, but it can take pressure differentials by acting as a support to pressure containing layers or other helix layers. As there are gaps between the individual wires, the adjoining layers must have sufficient local strength to bridge these gaps to avoid extrusion into the helix layer.

The air between wires does not transmit any forces, hence the wires are subject to a bi-axial stress condition consisting of axial stress and radial stress due to contact (see fig. 4). The difference between the forces acting on the inner and outer surfaces of the wire determines the axial stress through the hoop effect.

The stress-strain relationship of the non-bonded helix layer may be derived by:

* Formulation of Hooke's Law for bi-axial stress for the helix wire.

* Invert the equations and transform to average stresses.

* Transform from the spiral axis to pipe axis.

Note that the layer stiffness matrix in itself is singular, i.e. a simple helix layer may only take forces in conjunction with other layers (at least one layer with opposite lay angle).

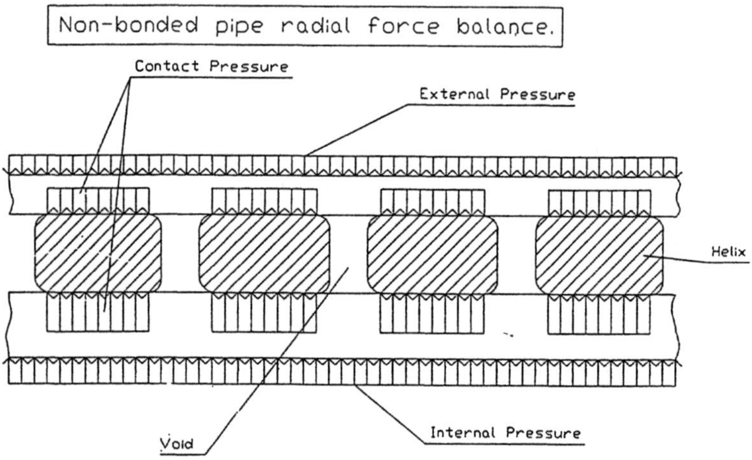

Figure 4 Non-bonded Helix Wire Layer

4.3 Bonded Helix Layer

The bonded helix layer resembles to a large extent the non-bonded, the only obvious difference is that the void between the wires is replaced by a rubber compound which is bonded to the wire material. As the rubber has very low stiffness compared to steel and even as compared to polyamide (about 1/80), it seems natural to assume that the effect of this rubber is negligible for axi-symmetrical loads, where strains and displacements are small.

The load transfer mechanisms are however quite different. As the pressure loads are transferred to a non-bonded helix layer by direct contact with a fairly stiff layer, in bonded pipes the helix layers are always separated by rubber adhesion layers of similar low stiffness, and there is consequently no distinct border between the rubber inside the bonded helix layer and the neighboring rubber layer. The rubber has a Poisson ratio close to 0.5, and with its low Young's modulus, it behaves almost as a fluid when subject to high pressures.

As the rubber layer is not stiff enough to transfer the pressure as contact loads on the helix wires, the rubber between the wires must resist extrusion by shear resistance, being bonded to the surface of the wire (see fig. 5). The limiting shear strength is a parameter which together with the tensile strength of the wire determines the pressure capacity of the helix layer.

As can be seen from the figure, the rubber between the wires will carry pressure, and due to its fluid-like behavior, it will apply a pressure load on the sides of the wires. Thus the wires will be subject to tri-axial stress, the pressure load normal to the wires being the third normal stress component, in addition to shear stresses. The main global effect of this is that a high lay-angle helix layer under pressure load will produce an axial compression force which has to be carried by the external layers with lower lay angles. There will thus be an effect of redistribution of loads from the high lay angle internal spiral layer to the external lower angle wire layers, as compared to a similar non-bonded model.

A narrow helix spacing increases the pressure containing capacity, but reduces the bending flexibility of the pipe. Thus for some high pressure designs, fabric layers are included to reduce the loading on the rubber between the wires. Depending on the effectiveness of this measure, the pressure distribution in the rubber will be somewhat modified, and more similar to the non-bonded pipe.

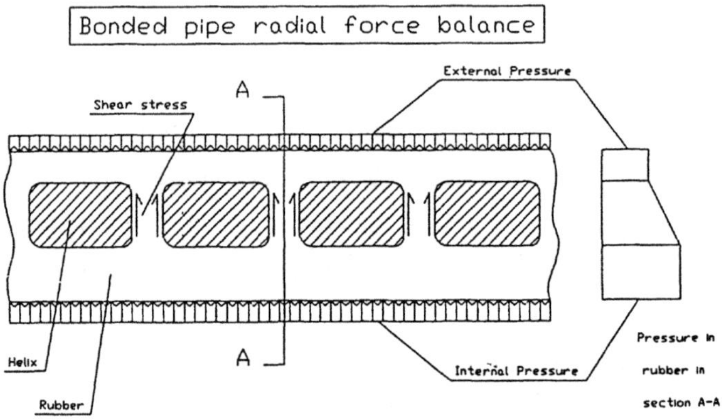

Figure 5 Bonded Helix Wire Layer

5 BONDED LAYER STIFFNESS FORMULATION

The method to derive the stiffness relations for the bonded helix layer is similar to the non-bonded, except that two materials are involved. the following assumptions are made:

* Both rubber and helix materials are isotropic.

* The rubber between helices transmits pressure forces through shear to the helix without significant radial deformation, i.e. the radial deflection of the rubber between helices does not significantly affect the radial deflection of the adjacent rubber layers (only the helix has an effect on the radial deflection).

First, Hookes Law for tri-axial stress in the two materials may be written.

The following indexes will be used:

r - rubber
s - spiral
- no index (r or s) means mean stress/strain

i - in direction of spiral
n - normal (transverse) to spiral
r - radially

i.e. ε_{is} is strain in the spiral along spiral, ε_r is mean stress in radial direction.

In direction of spiral (index i):

(1) $\varepsilon_{ir} = (\sigma_{ir}-\nu_r(\sigma_{nr}+\sigma_{rr}))/E_r$

(2) $\varepsilon_{is} = (\sigma_{is}-\nu_s(\sigma_{ns}+\sigma_{rs}))/E_s$

ν_r and ν_s are poisson ratio for the rubber and helix materials respectively.

Normal to (transverse) spiral (index n):

(3) $\varepsilon_{nr} = (\sigma_{nr}-\nu_r(\sigma_{ir}+\sigma_{rr}))/E_r$

(4) $\varepsilon_{ns} = (\sigma_{ns}-\nu_s(\sigma_{is}+\sigma_{rs}))/E_s$

Radially (index r):

(5) $\varepsilon_{rr} = (\sigma_{rr} - \nu_r(\sigma_{1r} + \sigma_{nr}))/E_r$

(6) $\varepsilon_{rs} = (\sigma_{rs} - \nu_s(\sigma_{1s} + \sigma_{ns}))/E_s$

Compatibility assumptions (see discussion above):

In direction of spiral, strain is equal for both spiral and rubber and stresses are "independent":

$\varepsilon_{1s} = \varepsilon_{1r} = \varepsilon_1$ and $\sigma_1 = A/at \cdot \sigma_{1s} + (1 - A/at)\sigma_{1r}$

where:

A = cross section area of the helix.

a = helix spacing (normal to helix).

t = thickness of layer.

Normal to spiral, the stress is equal for both spiral and rubber and strains are "independent":

$\varepsilon_n = A/at \cdot \varepsilon_{ns} + (1 - A/at)\varepsilon_{nr}$ and $\sigma_{ns} = \sigma_{nr} = \sigma_n$

Radially, constant pressure is assumed on rubber and spiral and only spiral can transmit displacement to other layers:

$\varepsilon_{rs} = \varepsilon_r$ and $\sigma_{rs} = \sigma_{rr} = \sigma_r$

The 12 equations above may be combined to obtain a relation between mean stresses and strains as referred to the direction of the spiral. Then the stresses and strains are transformed to mean stresses and strains in the pipe axis. Note that if any of the materials has a poisson ratio of 0.5, the matrix becomes undefined (similar to the simple isotropic material tube). The system of equations is however valid for all other combinations of material coefficients.

If the elastic modulus and poisson ratio of the rubber material is set to zero, the bonded model produces a layer stiffness matrix which is identical to the non-bonded model's. Similarly, when the material properties are set equal for the two materials, the model produces a layer stiffness matrix which is identical to the simple tube layer matrix. This indicates that the bonded model is theoretically consistent. The validity for a real case of steel and rubber is discussed below.

6 EXAMPLE, BONDED VS. NON-BONDED

6.1 Introduction

To show the effect of the bonded model, comparison between the non-bonded and the bonded models as implemented in the computer program FLEXPIPE is presented for a simple load case with 1430 bar internal pressure on a high pressure bonded pipe design. A summary of the construction data is shown in figure 6.

Note that for this load case, no negative contact pressure occurs in the pipe. This is convenient, as in this case it is easy to isolate the effect of the bonded stiffness formulation.

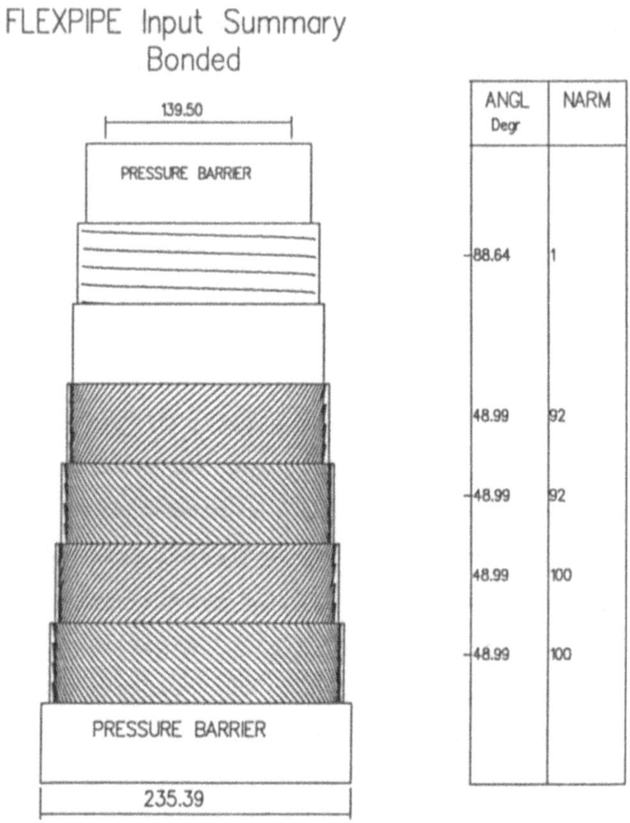

Figure 6 Bonded pipe description summary

6.2 Pressure Differentials

Computer plot no 1 in the end of this paper shows surface pressures for non-bonded and bonded cases. Note how for the bonded case, the 4 cord layers carry a larger part of the pressure differential. This is shown explicitly in the bar diagram in figure 8.

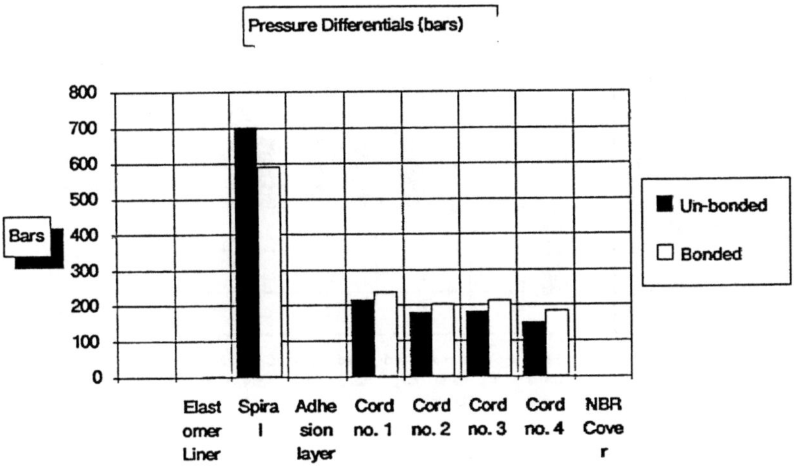

Figure 7 Pressure Differential Comparison

6.3 Radial Displacements

Computer plot no. 2 shows radial displacements for the two cases. Only minor differences are found.

6.4 Axial Tensions

Computer plot no. 3 shows axial tensions in each layer for the non-bonded and bonded cases. The sum of axial tensions for each layer balances the end-cap tension load which is given as AXTENS in upper left corner of each graph. Because of its low stiffness and poisson ratio close to 0.5, the inner rubber liner transmits an axial pressure load almost as if it were a pure fluid. The most apparent difference is that the inner spiral produces an end-cap tension force as well, due to the pressure in the rubber between the helices. This tension force has to be balanced by increased tension in the external steel cord layers. This redistribution is shown explicitly in figure 8.

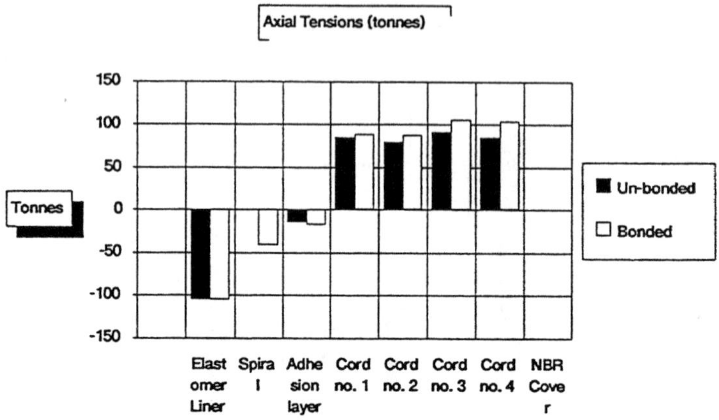

Figure 8 Axial Tensions Comparison

6.5 Stresses In Spirals

Computer plot no. 4 shows stresses in helices for the two cases. The redistribution of forces due to pressure loads is clearly apparent in the stress values as well, giving a marked decrease in the inner spiral stress, and corresponding increase in the external steel cord layers (figure 9).

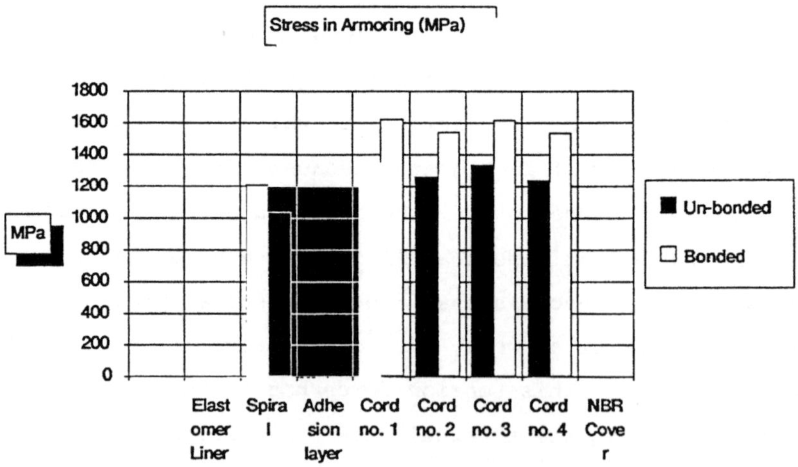

Figure 9 Stress Comparison

7 COMPARISON WITH TEST RESULTS

A series of burst pressure results has been made with the bonded pipe design described in chapter 6.1. In addition to recording burst pressure and failure mode, deflection parameters have been recorded. Based on tested tensile strength of the spiral and cord materials, burst pressure has been estimated by the FLEXPIPE computer program, using both the non-bonded and the bonded models. The results are compared in figure 10.

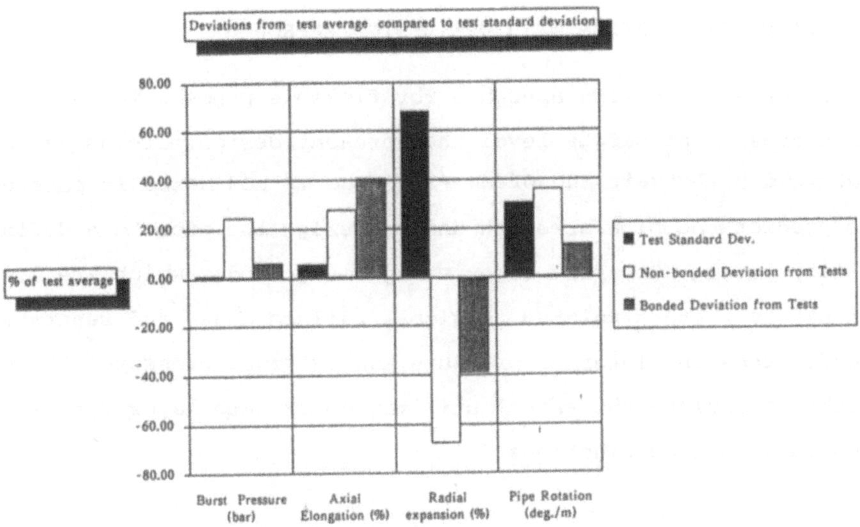

Figure 10 Comparison with test results

As can be seen, the bonded model predicts more accurately the burst pressure, as well as predicting the correct failure mode (rupture of the cord layer). The axial elongation is clearly over-estimated by both models and somewhat more by the bonded model. This deviation is probably due to the non-linear behavior of the pipe at these high strain levels, which is not accounted for by this linear model. Both radial expansion and pipe rotation predictions by FLEXPIPE are within the standard deviation of the test results, the bonded model seeming to give some improvement in accuracy. Note that the high percentage deviations for the radial expansion are somewhat misleading, as the mean measured test value here is small.

8 PIPE OPTIMIZATION WITH NEW DESIGN APPROACH

An alternative design approach for flexible pipes which gives a more consistent safetv level than present design methods has been outlined by Løtveit and Often /5/. The method makes it possible to predict the pipe strength and to design the pipe to a defined safety level. Contrary to existing design codes which are based on safety factors related to global failure modes and separate loads (tension, internal pressure and external pressure) the new method calculate the actual utilization of each layer for all possible load combinations.

For deep water where combined loads are important, pipe design based on tension, internal pressure and external pressure as separate loads may over or under estimate the layer utilization significantly resulting in too small safety margin or excessive strength.

The pipe strength governs both cost and weight of a deep water flexible pipe and may hence be used for optimization. Figure 11 shows the resulting safety margin if a deep water flexible pipe is optimized based on:

- This new design approach
- Typical oil company design specification
- Veritec JIP guidelines

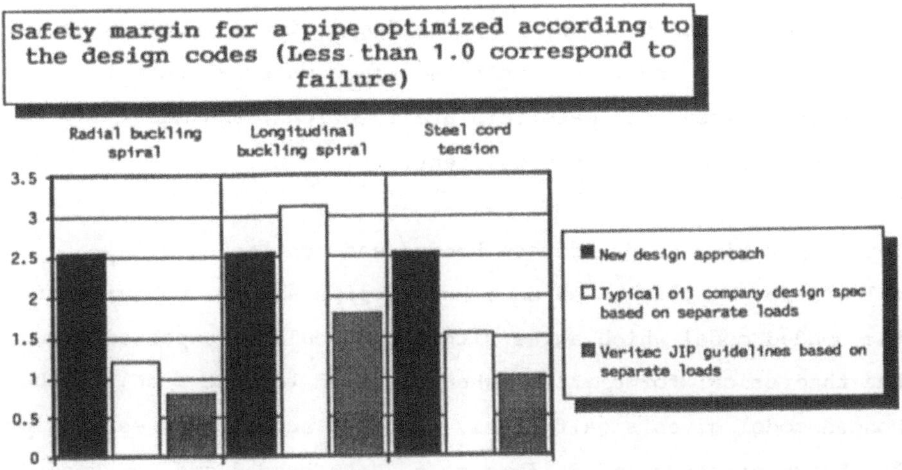

Figure 11 Safety margin for optimized pipes

The optimization is performed by reducing the armour layers strength until the loads from a case study (1000m depth and 3000psi) result in maximum permissible stress. The permissible stress varies between the design codes. The existing codes specify that combined loads shall be considered however no method for considering combined loads is outlined and the specified safety factors for separate loads are hence used for this optimization. Deviation between theoretical and real strength of the pipe is also covered by the new design approach.

As shown no or very small safety margin will be left if the optimization is performed with either the Veritec JIP guidelines or the typical oil company specification.

9 CONCLUSIONS

The bonded model which has been developed for axi-symmetrical loads, models in a somewhat simplified way the load transfer mechanism in a bonded pipe. As it degenerates to the standard non-bonded formulation when the stiffness of the rubber material is set to zero, and to a simple isotropic material tube layer formulation when both materials are identical, the model may be considered theoretically consistent.

The comparative study between bonded and non-bonded model calculation results based on a bonded pipe shows a behavior of the bonded model which agree with what should be expected from the theoretical formulation. When compared to test results, the bonded model gives significantly improved accuracy in estimating the burst pressure and failure mode. Both bonded and non-bonded models over-estimate axial elongation, the bonded model somewhat more. This is judged to be caused by the linear simplification in the model. For both radial expansion and pipe rotation the accuracy is improved for the bonded model, keeping well within the standard deviation of the tests. These tests, along with other tests not reported here, indicate that the proposed model provides an improved description of the load distribution in a bonded flexible pipe.

It can thus be established that the approach to model a bonded structure as an non-bonded, and to only allow for negative contact pressures between the layers, does not give the accuracy required. The detailed load transfer in a bonded structure must be considered.

As the same model may be used for both bonded and non-bonded pipes, it should be well suited for verification purposes, as it does not rely on empirical factors to model the difference in behavior of the two different types of flexible pipe.

The need for a consistent design approach taking proper account of layer utilization in combined load cases and considering all factors influencing on pipe strength is required for deep water and other new applications. The method outlined by Løtveit and Often /5/, combined with the theory presented will significantly improve the flexible pipe design process for both bonded and non-bonded pipes.

10 REFERENCES

/1/ Feret et Bournazel: "Calculation of stresses and slip in structural layers of unbonded flexible pipe."
ASME Transactions, reprinted August 1987, Vol. 109 Journal of OMAE.

/2/ Vetterman et Peuker: "Steel reinforced elastomer pipes. Design approach and performance characteristics."
NPF Flexible Pipe Technology Seminar, Oslo 1986.

/3/ McNamara et Harte: "Three dimensional analytical simulation of flexible pipe wall structure."
OMAE 89-744

/4/ Goto, Okamoto, Araki et Fuku: "Analytical study on the mechanical strength of flexible pipes."

/5/ Løtveit S.A. , Often O. , "Limit State Design Approach for Flexible Pipes", Norske Sivilingeniørers Forening seminar, Flexible pipes and hoses offshore, Trondheim, januar 8-10, 1989.

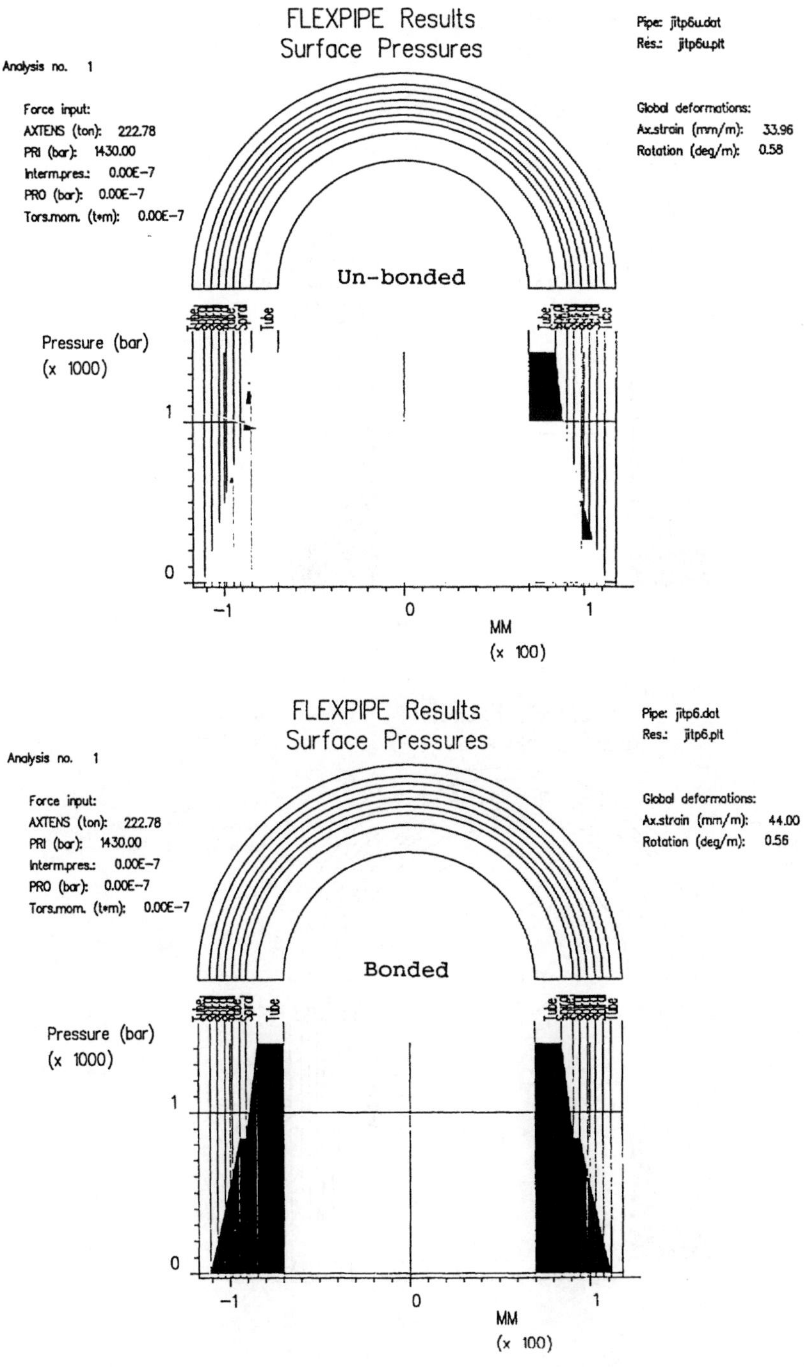

Computer plot no. 1 Surface Pressures

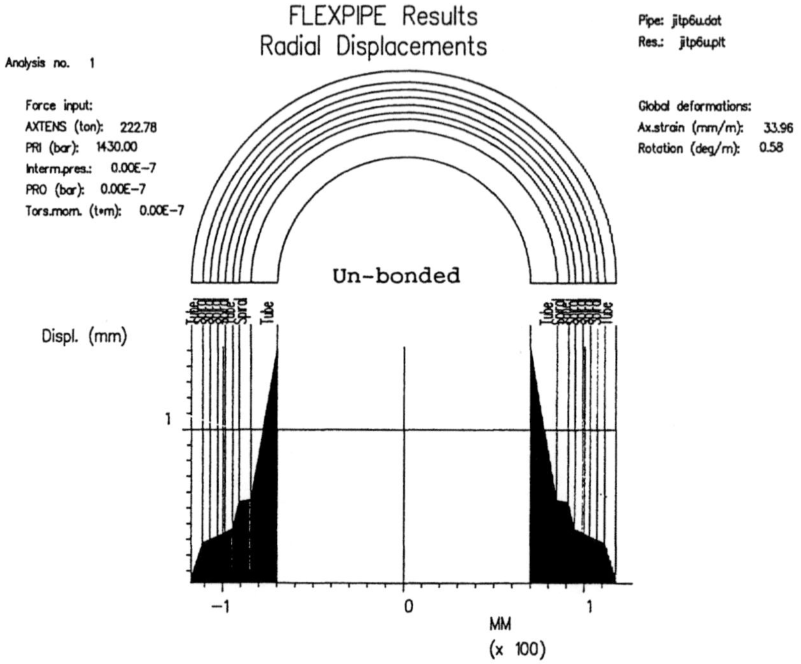

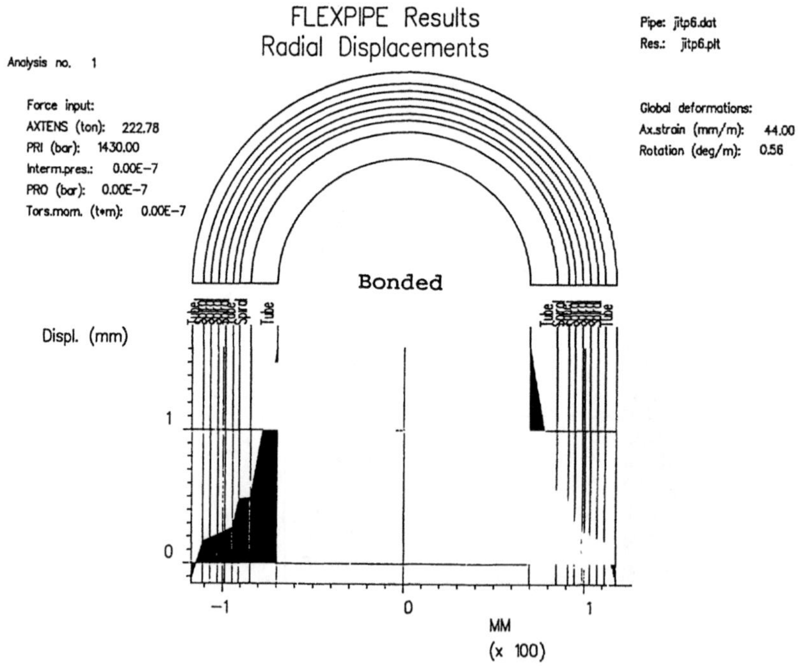

Computer plot no. 2 Radial Displacements

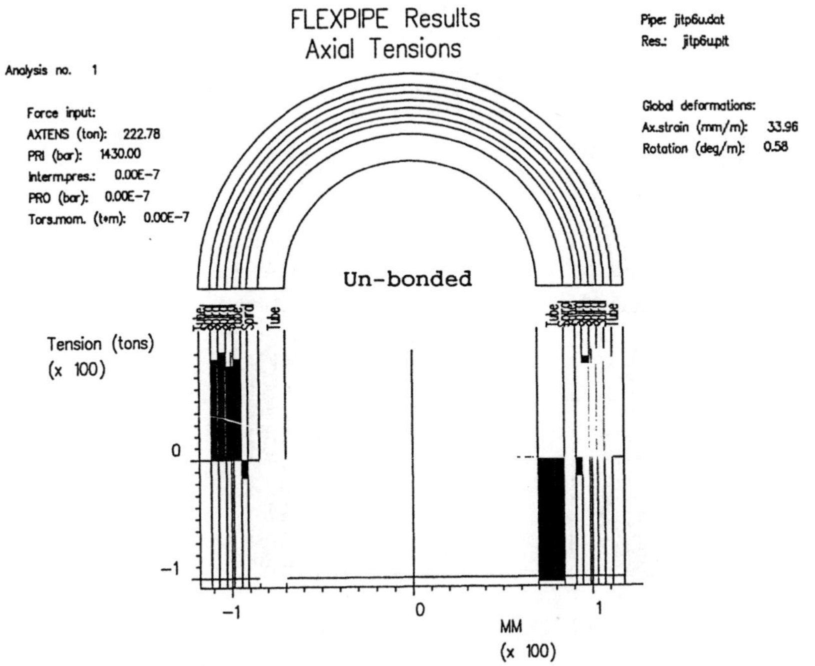

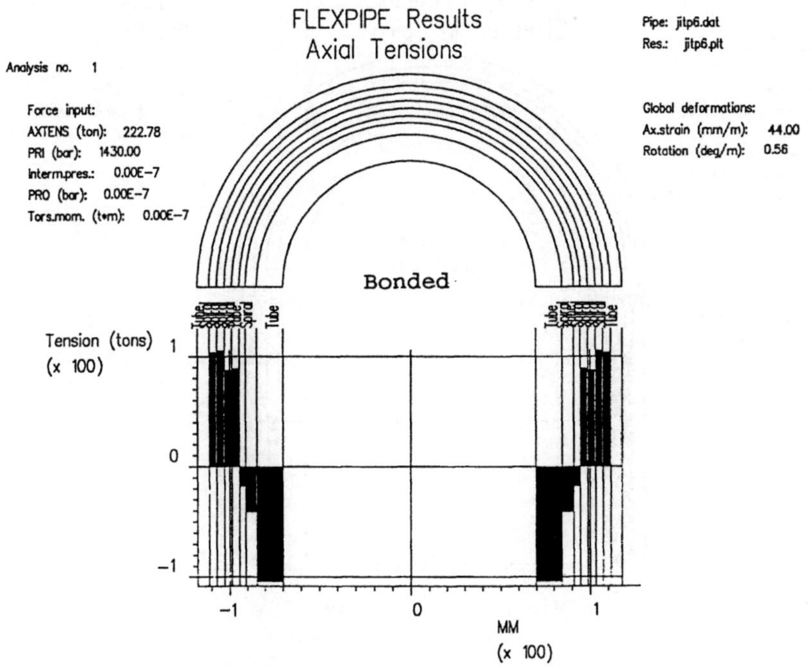

Computer plot no. 3 Axial Tensions

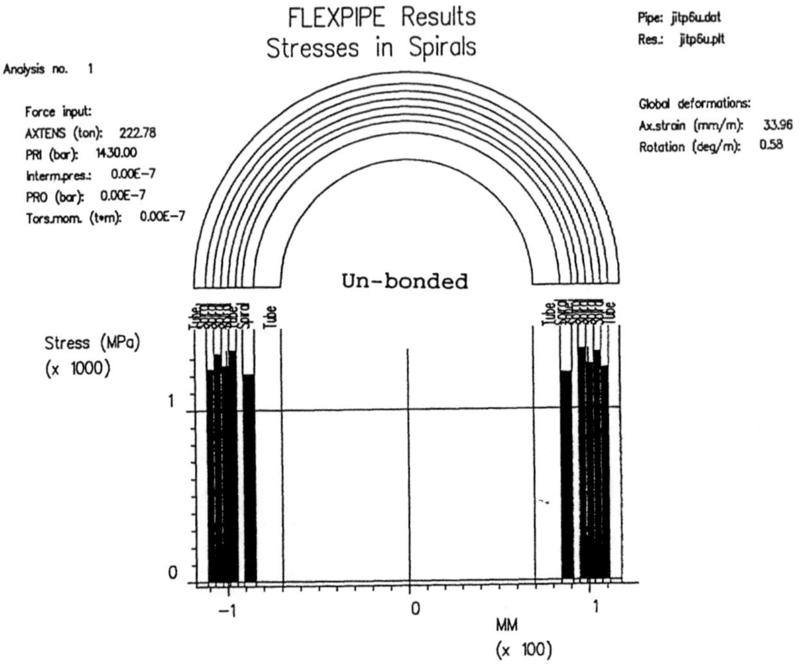

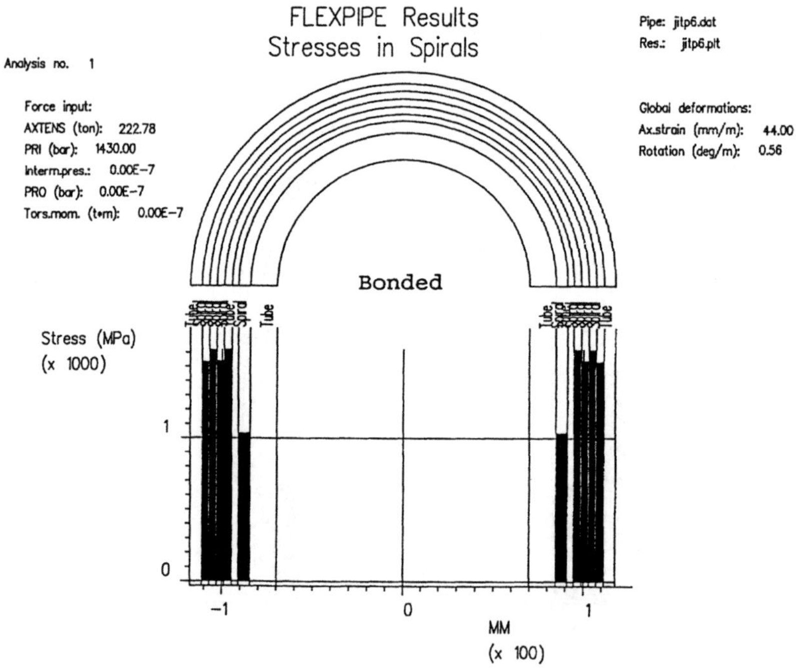

Computer plot no. 4 Stresses in helicies

ADVANCES IN FLEXIBLE PIPE DESIGN AND CONSTRUCTION

J.M. Neffgen
Neptun Pag-O-Flex A.S., Oslo, Norway

ABSTRACT

The industry´s strong desire during a time of uncertain oil prices to more economically exploit marginal oilfields has provided a strong stimulus to the development of associated pipeline technologies. Largely as a result of pioneering work performed in the 1970s, an alternative pipeline technology centered around the use of flexible steel reinforced pipe has been gaining greater prominence. The need for the introduction of a dynamic pipe capability into many subsea or floating prodution scenaria stems largely from the requirement to accomodate the system´s high inertial loads generated as a consequence of connecting fixed subsea hardware to a floating surface facility. Because of flexible pipe´s properties and composite construction, it provides a structure that can be quickly installed, variably configured, and which can adapt to a great range of system motions and environmental loads. Over the past 15 years, advances in oil resistant polymers and increases in the chemical resistivity of certain classes of steel have resulted in the construction of new types of flexible pipes capable of transporting ever harsher well fluids at higher temperatures. These and other improvements have resulted in greater product reliability and extended service lifetimes.

Volume 24: C. P. Ellinas (ed.), Advances in Subsea Pipeline Engineering and Technology, 111–143.

INTRODUCTION

Flexible steel reinforced pipe was first used in the 1940s to transport aromatics and fuels in support of the Normandy invasion, but only experienced its technical rebirth in the mid-1970s. The oil industry was then looking for a new pipeline construction which: could offer greater chemical resistance to transported fluids; could be quickly installed utilising less costly installation spreads; and which could inherently accomodate seabed undulations. The principal applications were as static seabed riser or flowline connecting adjacent fixed production platforms. The "flexibility aspect" the relatively low miminum bend radius (eight to ten times pipe i.d.) when compared to that of rigid steel pipe (up to 500 times i.d.) offered obvious installation benefits. The pipe´s ability to deflect without experiencing damage enabled relative large misalignments to be accomodated. At a time when a premium was demanded for the hiring of a dynamically positioned vessel from a relatively limited supply pool greater choice in selecting a simpler installation spread resulted in lower overall pipeline completion costs.

Early Uses of Flexible Pipe

The early uses of flexible pipe were confined to applications where the principal loads were imposed during installation rather than during long-term operation. Operating conditions were generally undemanding - pipes typically handled fluids up to 70°C at design pressures up to 200 barg with virtually no differential pressure across the pipe wall being experienced. Pig or TFL tool passage was unknown as was the concept of accomodating dynamic loads.

The principal concerns regarding the lifetime for these early flexible pipe designs were primarily related to the inner polymer liner's ability to resist ageing and degradation as a result of exposure to transported fluids. The secondary concern was its permeability with respect to associated gases. ED - Explosive Decompression was the catchphrase of the day. Concern centered around the fact that each flexible pipe initially utilised a polymeric inner liner material as a fluid conduit. These chemically-resistant materials although varying in permeability rates permitted gases (during pressurisation) to outwardly diffuse through or into the pipe wall as well as along interstitial spaces. In the event that the pipeline was suddenly shut-in and the pipeline consequently depressurised, entrapped gases would volumetrically expand exerting high internal forces at/in the liner. These forces could overcome their inherent shear strength and cause the polymers to blister or permanently deform.

Several different approaches by manufacturers towards resolving these problems were made in the early 1980s: improvements in the composition of the polymers (and or adhesives); introduction of a steel inner carcass or a gas-tight liner which could resist or completely prevent the pressure effects generated during ED; and the introduction of new ways of handling diffused gases utilising polymer materials designed to saturate and accomodate gases rather when internal pressures are altered.

Modern Pipe Designs

The 1980s saw a proliferation both in the number of pipe manufacturers as well as in the number of pipe designs. Due to the early successes of flexible pipe, the way was opened to expand the range and constructions of flexible pipe.

Pipes can today be generally grouped into two fundamental constructions - the "Bonded" and "Non-Bonded" types. The American Petroleum Institute (API RP17b, 1987) recently defined flexible pipe as being "...a composite of layered materials which form a pressure containing conduit. The pipe structure allows large deflections without a significant increase in bending stresses".

"Bonded pipes" are those in which layers are alternately applied and are chemically bonded to each other using bonding agents and special adhesives. Reinforcing steel wires and steel armour are applied to provide axial and radial strength. Steel components are kept separated by adding anti-extrusion (or anti-friction) textile reinforced fabric layers. Pressure distribution is consequently imposed and steel contact is avoided thus reducing internal heat build-up, wear, and fretting effects. Final "curing" of the pipe structure is achieved through a process known as vulcanisation (the application of heat and pressure).

"Non-bonded" pipes are also constructed using alternating layers of polymers, steel and flat wire reinforcement surrounded by tape or textile materials. The introduction of a lubicating medium or intermediate sheath is used to minimise steel-to-steel contact forces. In this construction greater importance is paid to stress levels in individual layers and adjacent cont pressures. Manufacture is generally based on a process used in power cable fabrication. Examples of these generic pipe constructions are given in Figure 1.

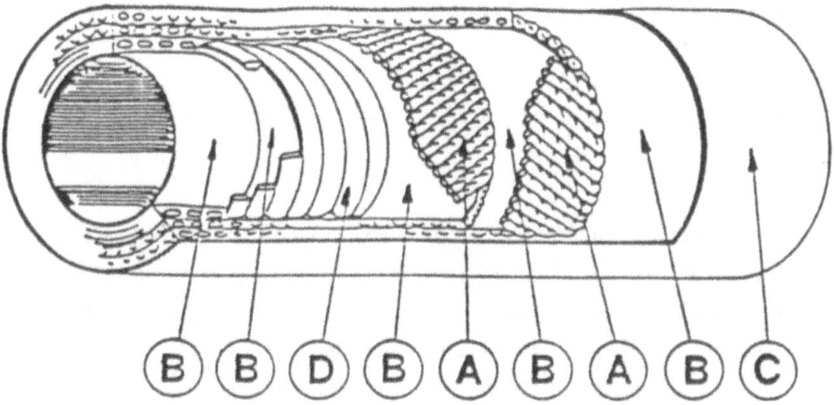

FIG. 1a: EXAMPLE OF A BONDED FLEXIBLE PIPE

KEY

A REINFORCEMENT WINDINGS

B FLUID CONTAINING LINER

C STRUCTURAL MEMBERS

D REINFORCING HELIX

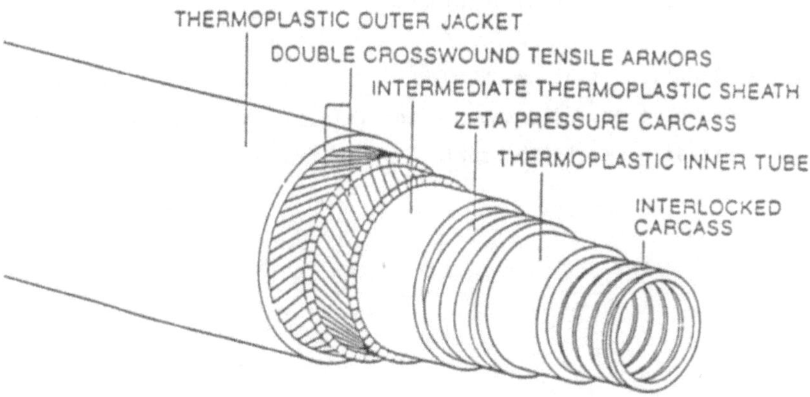

FIG. 1b: EXAMPLE OF AN UNBONDED PIPE

Development of Pipes for Dynamic Applications

The development of flexible pipe for use in dynamic applications constitutes the most significant milestone for this technology in the past decade.

The drive to develop a pipe structure offering excellent fatigue behaviour while also retaining sound mechanical properties stems from the need to reliably connect a fixed point on the seabed to a moving point which floats on an irregular and unstable sea surface. The principal difference between fixed and floating production is predominantly related to the system´s motions and the consequential high inertial loads. Flexible pipes whether for use as drooplines between a rigid riser to surface wellheads, or as catenary risers between the floating vessel and a fixed seabed structure, can be designed to connect out-of-planar elements without the pipes experiencing large deformations or significant reductions in lifetime. Figs. 2 and 3 illustrate two different examples of typical dynamic riser systems. In the Veslefrikk system (Fig. 1) a multi-riser system is laid out along a specially constructed riser balcony. The catenary risers are of relatively short-length (150m) but being in the splash zone are subjected to high environmental loadings.

In the second application, that of PETROBRAS´ Pirauna / Marumba production system, free hanging catenary risers are used (fig.2) to connect subsea wells to 411 m depth directly to the surface production unit.

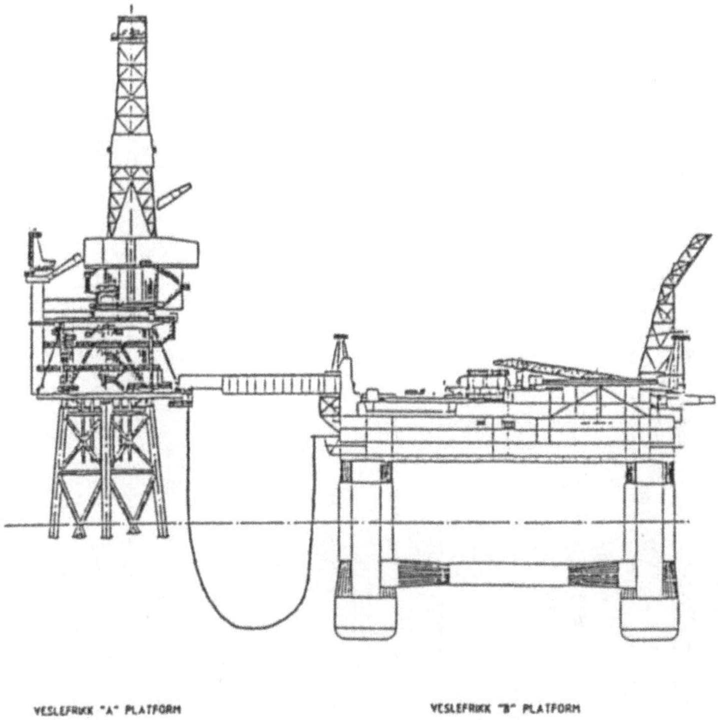

FIG. 2: Veslefrikk Catenary Risers in a Dynamic Application

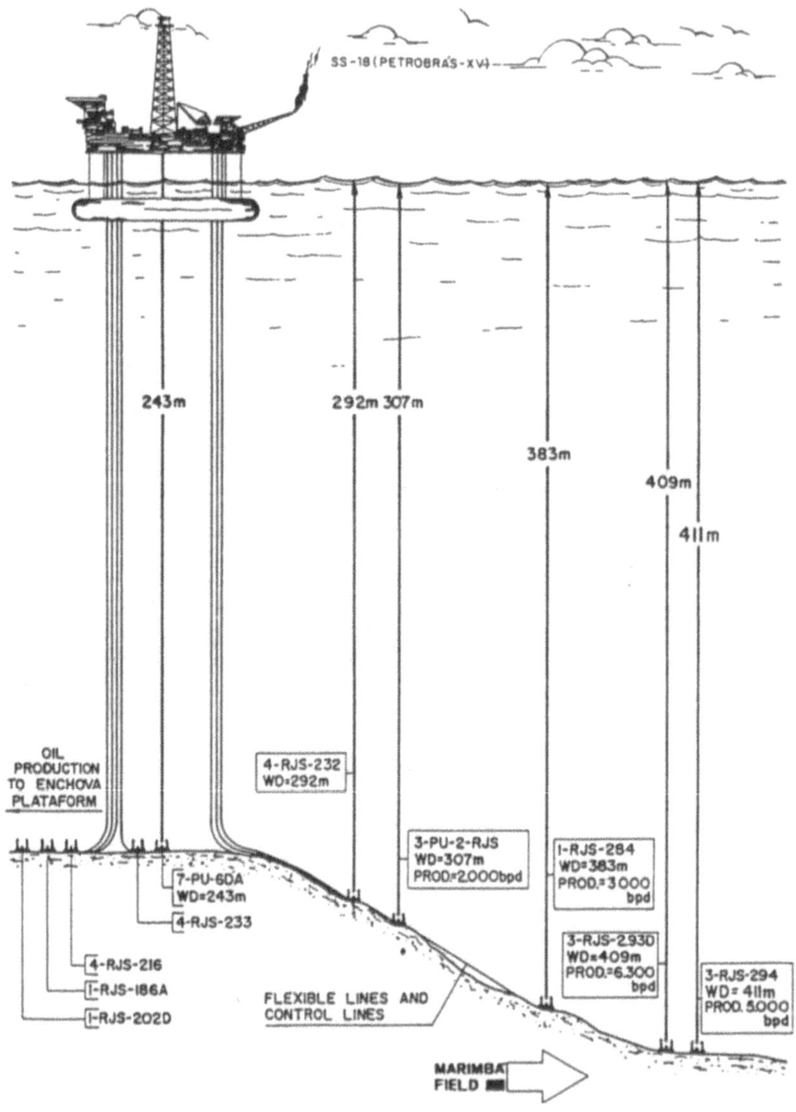

FIG. 3: Dynamic Flexible Riser System - PETROBRAS´ Marimba Field

ADVANCES IN FUNCTIONAL PIPE DESIGN

General Considerations

Due to its composite material construction, flexible pipe is able to utilise each component material in such a way that the cumulative effect of the many mechnical properties is greater than those contributed by any individual component. By embedding steel chord, for example, in an elastomer matrix one finds that its breaking capacity is considerably greater than that for steel chord alone. In another example, provision of fabric-impregnated layers between reinforcing steel (radial type) will substantially increase the ultimate pipe burst pressure capacity particularly during any bending. A similarly unreinforced pipe will burst at a lower pressure (PAG-O-FLEX Report -1987). Due to the use of polymers in pipe construction, flexibles will also have higher damping and thermal coefficients than will rigid pipe. These properties plus the method of internal reinforcement provide flexible pipe with a significantly greater resistance to buckling or impact damage than that offered by rigid pipe.

Another interesting feature is that due to the nature of the fabrication process design changes can be incorporated in the pipe structure almost up to the eleventh hour. The manufacturing process (either mandrel-based or rotating head based) is a step-by-step one which will largely be reversible. Pipes are generally constructed either on a rotating steel mandrel, or they are fabricated by extruding polymers along a central steel structure around which armouring equipment will rotate. In both cases, the process offers considerable flexibility to incorporate additional layers, microstructures, or even fibre optic cables within the pipe body (Neffgen, 1988). However, to obtain a clearer understanding of the nature of a flexible pipe design, it is perhaps necessary to understand the functions of some of its key layers.

Inner Liner

Beginning with the pipe's internal liner, material selection must be of paramount concern. Chemical rather than mechanical considerations are most important because the internal liner largely serves as a simple fluid conduit through which well (or exported) fluids will pass. The liner selection is governed by the desired chemical and corrosion resistivities and the presence of free gases (Methane, Ethane, CO_2, H_2S, et. al.). As a general guideline, polymeric materials are utilised in gas-free applications or those in which high gas partial pressures are absent. Typical applications are: methanol or dead crude transport; water or chemical injection operations; acidization; drilling; and some well test operations. When gases are present or their partial pressure is high, a stainless steel carcass or corrugated steel inner liner is substituted in order to reduce or even prevent gas diffusion and provide a strong shell structure capable of resisting Explosive Decompression effects.

The selection of polymers from those generally available can be a daunting one and the choice is made only after consultation with chemists and specialist polymer suppliers. Polymer liners are generally made from extruded thermoplastic material (Nylons, Polyamides, or Polyethylenes) or layered elastomeric material (generally Nitrile Rubbers, Saturated Nitriles, or Polyolefin Rubbers). A new class of Thermoplastic Elastomers is presently being studied. These new materials offer good oil and pressure resistances and are currently in use as seal rings for high pressure application.

Steel materials used for the inner carcass or corrugated tube have been considerably improved over the past decade. Today exploitation is moving towards higher temperature applications (90°C or greater). In view of this fact, there is a need to maintain structural strength while improving chloride resistance. Because of these facts, there is an industry desire to move away from ordinary stainless materials, such as: 304 and 316L and go towards the Austenitic and Duplex steels (6% Mo and 21% Cr content). These materials have been in use for several years in a range of applications and temperatures varying from 80° C to 110°C (Neffgen, 1989).

Carcasses or Liners are made by cold deforming thin, flat steel strip. Considerable attention must therefore be paid to Rockwell Hardness values and locked-in stresses. According to AVESTA Stahl of Sweden (Wallen and Abrahmsen, 1984) due to the shape (interlocked or corrugated wave) of the inner liner, and because of the forming process localised rather global corrosion attack presents the primary concern. The importance of the inner carcass integrity will also increase not only because of chemical resistance, but also in structural importance with increasing water depths. In some pipe designs, the carcass itself provides the principal reinforcing method to resist hydrostatic collapse. In this case stress corrosion cracking may play a more important consideration than previously thought.

Reinforcement and Armour

Flexible pipe is generically called "flexible steel reinforced pipe" due to the presence of large quantities of reinforcing steel in the pipe structure. Axial reinforcement is provided by the diagonal application (generally at 54° from the neutral angle) of cross-wound steel chords or flat steel tendons and is assisted by the helically wound armour. The number of steel layers will vary from two to six depending on internal pressure and axial pull requirements. Radial reinforcement is provided either by a) the inner carcass itself, b) by a helically wound steel spiral, or c) via the provision of an additional Zeta-type wire or carcass.

Improvements have been made in the past decade in the batch testing methods to control quality of manufacture. Steel chord wires are today coated in brass or aluminium to provide a better mechanical key into the surrounding elastomer matrices. The methods for cold forming of the liner or carcass are today more controllable than previously resulting in reduced residual stresses.

Figures 4 and 5 illustrate the differences between the types of steel liner and reinforcement used in pipe constructions.

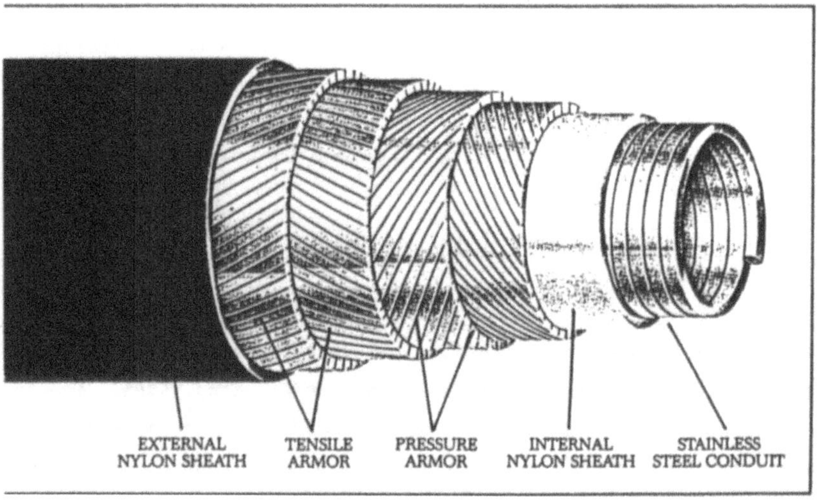

FIG. 4: Steel Carcass Type Pipe with Flat Tendons

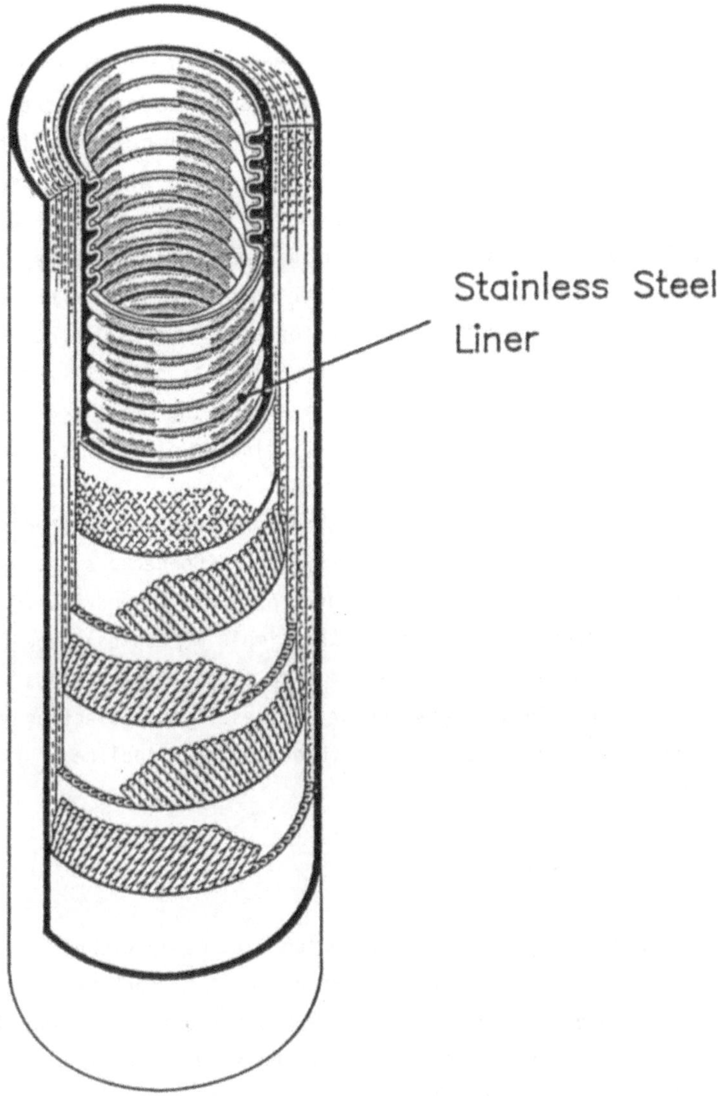

FIG. 5: Pag-O-Flex Gastight Pipe with Corrugated Liner

End Fittings and Connectors

To terminate the pipe reinforcement and transmit internal pipe loads end fittings are used. These end fittings are made from steel, although carbon fibre has been considered. They are either of the "built-in" type-meaning that they are constructed having a series of steps or teeth which permit the steel reinforcement to mechanically key into the end fitting detail so that the termination becomes an integral whole; or, they are of the "post-mounted" type. In the latter construction, the end fitting is mounted either by splicing the pipe reinforcement materials onto the end fitting following pipe manufacture; or, it is squeezed or crimped-on using a hydraulic crimping device with final sealing being achieved via injection of a resinous material or by the inclusion of mechanical seal ring(s).

End connectors refer to the method by which a terminated pipe can couple to an adjacent pipeline or piping system. Connectors are made from steel and are either machined as an integral part of the end fitting or are welded to it. Connectors can be of the simple mechanical type (flange, union, or hub) or can be of any number of more sophisticated types of hydraulically actuated couplers or disconnect devices. Their selection will depend on the connection philosophy used throughout a pipeline or floating production system.

External Coverings

The pipe body will be protected externally either by applying a durable textile material as a tape layer around it, or by extruding a polymer sheath over it. Depending on the requirements to resist abrasion, chafing, or mechanical fretting, the wear mode will govern the final material selection and layer thickness. In cases where extreme fretting or chafing damage is likely, a slip-on sleeve type mesh (Chinese Finger) or external interlocked steel carcass can also be added. Fire resistance will be provided (today to Lloyds Bulletin, i.e., 700°C for 30 minutes while the pipe is fluid-filled and under pressure) by providing an additional external textile covering made from glass fibre or ceramic impregnated materials.

Evolution in Design Considerations

Current knowledge of flexible pipe rests largely on results from R&D investigations and input from field experience gained over the past 15 years. Pipe technology had previously centered around a basic understanding of the functions of the individual pipe layers and assessment of their interrelationship. Nunmerical predictions were expressed in relation to the importance of individual layer stresses. Of particular interest had been the correlation of test results gained from component tests with those results gained from testing sample pipe structures. Very often the only S-N curves available were derived as a result of component tests. Facilities and rigs did not exist to duplicate loads on pipe structures and such testing to be meaningful was likely to be costly and take a long time to perform.

Due to the composite nature of its construction, flexible pipe often confounds clear analysis as it exhibits a heterogeneous mechanical behaviour unlike that of steel. Because of changes in material composition in individual layers forces generally do not act isotropically but exhibit a complex behaviour making stress analysis difficult.

This revised understanding strongly influences today's analytical efforts and is supported by information obtained from:

a) Component and laboratory tests; b) static and dynamic mechanical tests; c) scale model or basin tests; d) by computer modelling; and e) from a review of operational experience.

Another lesson learned from such experience has been to remove the commonly held belief that flexible pipe is a "black box technology". Today a better understanding of pipe behaviour can be derived from a multitude of sources as well as from long-term fatigue test results, motion response data, and results from long-term chemical exposure of polymers. Such data can be verified by several finite element programs.

PAG-O-FLEX, one manufacturer of bonded flexibles has carried out long-term fatigue tests during a 2.5 year joint industry program funded by the German Science Ministry and six major oil companies (see Kokkenowrachos, Peuker, Giese OTC, 1987). As a result of tests made with a number of 6" x 6000 psi riser pipes (tested without failure over more than 20 million cycles in dynamic bending and tension at simulated 100 yr. load conditions), they have been able to conclude that wear not fatigue is likely to constitute the primary mode of failure. A synopsis of the relevant test data and load conditions is presented in Tables 1 and 2. As a result of these tests at varying load conditons over a great number of cycles, the researchers were able to obtain useful data related to pipe ultimate capacities and relate this data to S-N curves formulated as a result of component testing.

SHELL International and BRITISH GAS have also in the past 5 years taken mechanical scale testing one step still further and have developed a new kind of "floating" test rig capable of simulating dynamic riser laods. As a result of such testing, long pipe lengths can be tested over an accelerated lifetime. The imposed loads can then be accurately related to the numbers of cycles to failure.

Today less emphasis is placed on layer stresses largely due to the difficulty in measuring such stresses and in numerically isolating them in relation to their unisotropic distribution in surrounding layers. Manufacturers today use in-house stress analysis software which is continually revalidated by comparing measured burst pressure results obtained from destructive tests against theoretical figures. In a recent study by BrasNor (O. Often, S.A. Loetveit, 1989) some 80 different burst pressure results were compared against computer predictions made using PAG-O-FLEX's FE program "Aufbau". The comparison showed that the spread of measured vs. actual test results were 90 percent of the time within close correlation thus justifying a high confidence in the numerical methods used.

TABLE 1: TYPICAL VALUES FOR A 6" X 6000 PSI BONDED FLEXIBLE RISER PIPE FOR 350M DEPTH

TYPE OF TEST	MINIMUM INDUSTRY VALUE	CALCULATED VALUE	MEASURED VALUE
Burst pressure (psi)	18,000	20,700	20,600
Collapse pressure (psi)	520	1,800	2,145
Tensile strength (kN)	200	1,450	1,450
Bending stiffness (Lb. x in^2)	None	1×10^6 (0 psi) 1×10^6 (6000 psi)	$1,4 \times 10^6$ $6,3 \times 10^6$
Tension stiffness	None	3×10^8	$3,1 \times 10^8$

TABLE 2: ABSTRACT OF DYNAMIC TEST RESULTS FOR 6" X 6000 PSI BONDED RISER PIPES

TEST	LOAD CONDITIONS	PIPE I.D. (MM)	NO. OF CYCLES WITHOUT FAILURE (MILLION)
Axial Pull	mean load 200 kN int. pres. 414 Barg temp 80°C	140	11.9 m at 200+/- 160 kN 11.2 m at 200+/- 100 kN 15.5 m at 200+/- 60 kN 23.6 m at 200+/- 30 kN
Bending	various radii 2.5 - 7.5 m temp 80°C int. pres. 414 b	140	21.3 m at 19.5 m

Notes: Testing conditions cover worst case laods for North Sea for 100 yr. condition

Strain gauging has also been used recently inside a bonded pipe at the interface between the corrugated inner steel liner and first adhesion layer. This work was carried out as part of a 1-year joint industry program to dynamically test gas risers which were sponsored by three oil majors. In this program, it was found that the measured strains corresponded well with computer predictions concluding that layer stresses can be accurately forecast. The use of fibre optic cables as gauges to measure strains in areas of microbending is a technique pioneered in the USA. This technique as applied to flexibles may offer still further possibilities to improve data gathering and validate numerical predictions particularly in regions of highest stresses. (McKeehan, D. et. al., OTC 5119, 1986).

Due to improvements in modelling software, a method for assessing ultimate capacities for both bonded and non-bonded pipes is available. Work in this area is presently being carried out by Norway´s SINTEF foundation in their Structural Division and by BRASNOR AS, a Norwegian subsea engineering contractor. A report on this work is presented in this book under the title of "Increased Reliability through a Unified Analysis Tool for Bonded and Unbonded Pipe" (O. Often, S.A. Loetveit, 1989).

Lastly, due to the expansion in the use of flexible pipes new research in the area of non-linear analysis have been undertaken largely as a result of institutional efforts. In a recent paper (Boef, Lange, Van den Boom, London, Dec. 1989) concerning the analysis of riser systems some new work was performed to verify pipe performance related to extreme conditions and fatigue. In a project known as DYNFLX carried out between 1983-86, Dutch researchers used model tests to gain an insight into hydrodynamic co-efficients and performed analyses based on the lumped-mass discretisation technique. To assess large scale deflections under combined bending and torsion, special scale model tests with PVC rather than flexible pipe were performed. Results showed that "...a proper description of the torsion of a riser could be achieved to accurately predict the structure´s response under conditions in which three-dimensional bending occurs".

ADVANCES IN MONITORING AND RELIABILITY ASSESSMENT

Evolution in design thinking has also been accompanied by an evolution in thinking concerning reliability of flexible pipe. It is said that the reliability of a component part will determine the overall reliability of the whole. Robert Ross in a recent paper (Ross,R., SPE 19273, 1989) as a result of his investigation of failures stated that "Components fail because they have some defects in design, material, or workmanship. Importantly, he pointed out that "... in many instances components can (however) operate successfully for lengthy periods (even having) relatively massive defects."

In two recent studies carried out on the behalf of two oil majors on the integrity monitoring of flexible pipes (Pag-O-Flex, confidential report, May 1988). A series of failure mode and effect analyses for both bonded and non-bonded pipes were performed. The principal aim of these analyses was to determine the significance of probable defects in relation to their capability to propogate and lead to failure. Likely failure modes were identified and related to failure mechanisms. VERITEC (Veritec Guidelines, JIP/GFP-02, 1987) has previously identified, as a first step, the most probable failure modes for flexible pipe as being:

* disbondment of bonded components,
* fretting or internal wear,
* corrosion of steel components,
* fatigue failure of component parts or structure.

Advances in Monitoring and Testing of Flexible Pipe

Reliability of flexible pipe must also be assessed like any other material using results gained from a suitably designed set of inspection tools. These tools must identify and quantify pipe defects both during manufacture and in service. As a follow-up, investigations must also be performed to classify defects and then relate them back to appropriate failure modes. An important fact to be highlighted with regards to reliability of flexible pipe (see Neffgen, J.M., Pipes Pipelines International, May 1988) is that due to its composite material construction a high degree of redundancy exists. Safety factors used are generally at least as great as those used for steel pipe. As examples, when comparing a flexible pipe having a design pressure of 350 barg or less it is usual to apply a 3:1 ratio (burst-to-design pressure). When designing to accommodate axial loads, the safety factor used will be as great as that for wire rope, typically 4:1 or 5:1 (breaking-to-allowable pull).

The method of pipe construction, the multi-layer composition, and the contribution made to individual component capacities from embedding components in polymer matrices, all assure that flexible pipe will have a favourable mode of failure. This implies that for a properly designed pipe the likelihood for sudden or catestrophic failure will be of very low order.

Inspection of flexible pipe is another area where related technological advances will have an impact on assessing pipe reliability. Following early work in this area by the pipe manufacturers themselves (Neffgen, J.M., Integrity Monitoring 1988), a new joint industry program co-ordinated by SINTEF of Norway as part of the "Floating Production Systems 2000" is currently underway. This 3-year industry wide program intends to progress pipe knowledge over four main areas:

* An understanding of pipe designs and methodology,

* An evaluation of test methods currently used with a view towards their standardisation,

* An evaluation of tools and techniques for the inspection of flexible pipe,

* A review of installation criteria and methodology.

In the area of non-destructive testing of pipes what can be concluded is that by adapting currently available NDT tools and training skilled operators, the industry will shortly have suitable methods for determining pipe integrity. Some promising NDT methods investigated are listed herein:

* Thermography and thermal imaging (topsides pipes),

* Real-time radiography,

* B and C-Scan ultrasonics,

* Acoustic emission,

* Use of radioactive isotopes for leak tracing

* Impedence Measurements.

ADVANCES IN SYSTEM TECHNOLOGY

Not only has attention been paid during the past decade towards general pipe improvements, such as: component reliability, and improved accuracy of design software; but considerable work has also begun towards the recognition of flexible pipe´s importance as a "key" mechanical subsystem in any production system. As products become more reliable and quality control procedures more routine, end users are shifting their emphasis towards the application end of pipe technology rather than on the product end. In this context, operators at an early stage need to work with manufacturers and should consider not only the design of the pipe itself, but also other factors, such as (VERITEC, JIP/GFP-02,1987):

* compatibility with regards to interfaces,
* functionability requirements,
* possibilities for in-service inspection,
* maintenance and/or repair requirements
* methods for detecting defect growth or propagation.

In applying these considerations to the design of an overall floating or subsea production system, a more detailed knowledge of pipe capabilities must be related to the constraints placed on it by: the environment, the individual system motions, and the reliability of any auxillary items or appurtenances (such as: buoyancy devices, bend limitors, hydraulic end connectors, and fixation structures).

Advances in Riser Pipe Construction

The past 10 years has seen the maturing in the industry´s understanding of pipe construction as well as the classification of pipes into two definite categories, Bonded and Non-Bonded. This fundamental understanding has lead to new confidence levels being reached. During the past 10 years a proliferation in the types of floating production equipment has stimulated pipe development specifically related to risers. Some of these are:

* Development of multi-bore pipes which can be connected to a multiple surface mounted hydraulic connectors.

* Development of a multi-function bonded type riser having fibre optic signal and electrical power cables incorporated into the structure (see - Sterzenbach, SOLS 1989).

* Development of light-weight glass fibres as axial reinforcement for risers intended for deepwater (600 m and below).

* Extension of testing work on large-bore dynamic risers (generally for export) to 20 inch i.d. and increases in their design pressures.

* Development of special high temperature service pipes using multiple layers of thermoplastic sheathes; or by incorporating special hydrogenated elastomers into bonded pipe´s structure to offer increased thermal resistance at elevated temperatures (from 90º C to 140º C).

* Improvements in altering pipe sectional properties and the introduction of increased localised weight into lower bend areas (bonded pipes). This feature can be used to significantly improve distribution of weight and improve on-bottom stability by maintaining similar unit weight to O.D. ratios for multi-riser systems (Neffgen, J.M., Optimisation, 1989).

* Introduction of anti-fouling compounds into the pipe structure to reduce growth by dense marine organisms.

<u>Advances in Riser System Technology</u>

The 1980s has seen the real increase in both the number and complexity of dynamic riser systems principally outside of Brazilian waters. Early Brazilian riser systems were less complicated than those on the North Sea and were installed in shallow waters (less than 280 m) using the free hanging configuration.

Over the past 10 years, catenary riser systems have been or will be shortly installed in most offshore sectors around the world at water depths ranging from 70 m (in the Far East) to 350 m in the Gulf of Mexico (Conoco's Green Canyon). The next steps into deeper water will be on pioneering projects such as SAGA's Snorre Riser system and PETROBRÁS' Marimbá and Marlim floating production units offshore Brazil. These risers are to be installed at depths of 350, 430, and 600 m respectively. Such sophisticated multi-riser systems are most certain to accelerate the overall pace for using flexible pipe in floating production developments. Major concerns remain, however, in operating in such deep water. These concerns can be summarised as follows although work has already begun in many areas:

Deep water concerns

* high crushing loads on pipes during installation,
* need for high hydrostatic collapse resistance,
* need to reduce top tensions and pipe weight,
* need for specialised installation equipment,
* need to investigate effects of slug flow and hydrate build-up caused by low temperatures and increased pipe length,
* need to investigate hydrodynamic effects and those induced by deep ocean currents.

Advances in Connectors and Bend Limitor Technology

No riser system would be reliable if it could not be designed so as to resist the anticipated and worst case loads or if it could not be readily installed or retrieved. The importance of bending to flexible pipe integrity should therefore be highlighted.

The need for safeguarding a safe minimum bend radius (MBR) in all design cases must herein be stressed because of its importance towards maintaining pipe serviceability. Perhaps it is worth restating the definition of minimum bend radius (VERITEC, 1987) as follows: "The minimum radius to which a pipe may be bent without damage, including temporary excessive ovality...while considering both short and long term effects on pipe and pipe materials." Irrespective of the desirable effect a low bend radius may have on a production system, the consequence of exceeding the MBR in any load case could result in permanent deformations within the structure, disbondment of internal layers, or, ultimately, collapse of the entire structure. This consequence would seriously hamper production operations and even system viability.

One benefit derived from the developments in analysis software is the industry's current ability to more accurately quantify bending moments and identify key points along the pipe axis where forces will be at their greatest. Generally, bending stresses will be highest at riser terminations and at areas of lowest bend radius, such as: areas where pipe lay over buoyancy devices or steel arches; at pipe touch down points; and at connections to seabed hardware and/or topside piping.

The use of bend limitors (either external restrictors or built-in stiffeners) at such areas allows bending moments to be safely accomodated and corresponding stresses to be redistributed. Today such limitors can be of several basic types as follows: Strap-on mechanical, Slip-on elastic, or Built-in elastic types. The latter Built-in (or integral) bend stiffener has been successfully used on the SOLS 8 inch riser (Sterzenbach, 1989) and on end terminations of Statoil's Veslefrikk risers. (Neffgen, J., Optimisation, 1989). Fig. 6 illustrates the (3) types of limitor and shows a comparison between stiffness and unit length

Some advantages of this type of steel reinforced built-in stiffener over strapped or slip-on bend limitors can be summarised as follows:

* shorter length (25-40% of previous length) however having a similar contribution to pipe stiffness,
* minimal (up to 75%) lower change in pipe o.d. thus reducing drag forces,
* low probability of slipping because of being mechanically integral with the pipe structure,
* reduced cost (40-50% lower cost than for the other types).

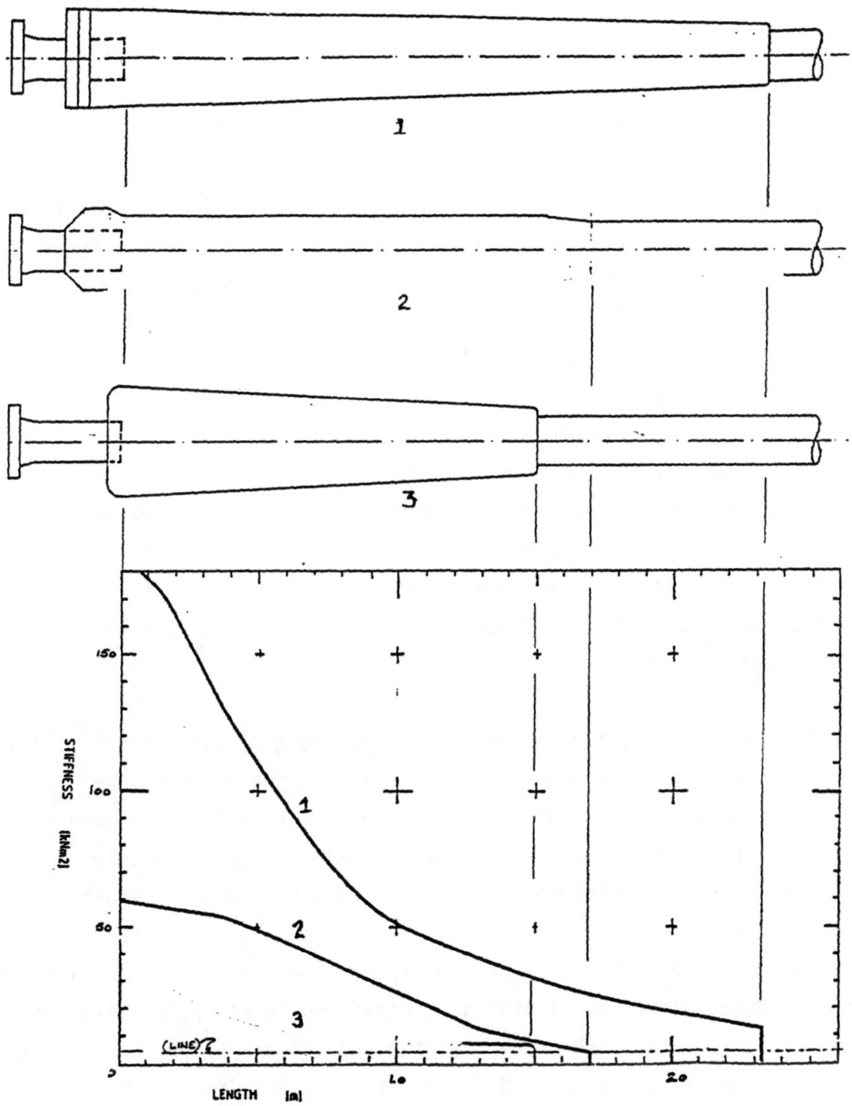

Fig. 6: Comparison between 3 types of bend limitor

In order to terminate the flexible pipe and to transmit pipe loads to more permanent structures, a variety of end connectors can be used. A basic decision regarding the design of a riser system is to decide on what philosophy and criteria shall be used to connect/disconnect the risers. A simple choice is to use a connector generally of the hub or conventional flange type. These connectors, however, are permanently secured via bolts or by using an external clamping device meaning that they cannot be simply or quickly disconnected without manual intervention. More sophisticated connectors are generally of the hydraulically actuated quick connect/ disconnect type (QCDC) or emergency release type (ERC). This latter type as shown in Figure 7 offers considerable advantages over the former QCDC type. The QCDC type had previously been used on PETROBRÁS early floating production systems as it was the only suitable connector available. The connector assembly weighed some 4.5 tonnes and required some 2.5 diameter space on a riser balcony. Ball valves were previously incorporated in the overall connector design in order to isolate pipe contents in the case of a sudden or emergency disconnection of the riser pipes. As a result of policy reviews into the need for disconnection, design changes have brought about an improved connector.

The ERC type connector relies on the priciple of disconnection only and operates similarly to the principle used for mating a diving bell to its deck mating chamber. This principle has successfully been used for many years. An ERC is a smaller, lighter weight (approximately 1 tonne) mechanical device which has no bell housing thus requires smaller installation space. No ball valves are provided because: a) the probability of disconnect is considered too low, and b) the pollution risk from entrapped fluids is limited and considered to be acceptable. In the case of the Veslefrikk riser system, all risers except the gangway mud transfer jumpers have been so designed as to remain connected during all operational cases, even when the vessel´s is in standoff position(s) thus reducing the likelihood of disconnects to a low order of magnitude. By carefully reviewing their safety design, Statoil were able to simplify their baloncy design through the use of simpler connector mechanisms.

Connection technology whether related to flowlines or risers is, however, another relatively young technology, but one where several decades of related work on pipeline connectors has already been performed. When considering that the connector must withstand all specific loads experienced by the flexible pipe including those from installation, operation, and disconnection its design must today heavily rely on experience gained from another related technology that of marine drilling risers. Experience with such dynamic riser operations can provide a significant input into the effects of lifetime for dynamically loaded end connections. Work in this area has already been started by a new API committee.

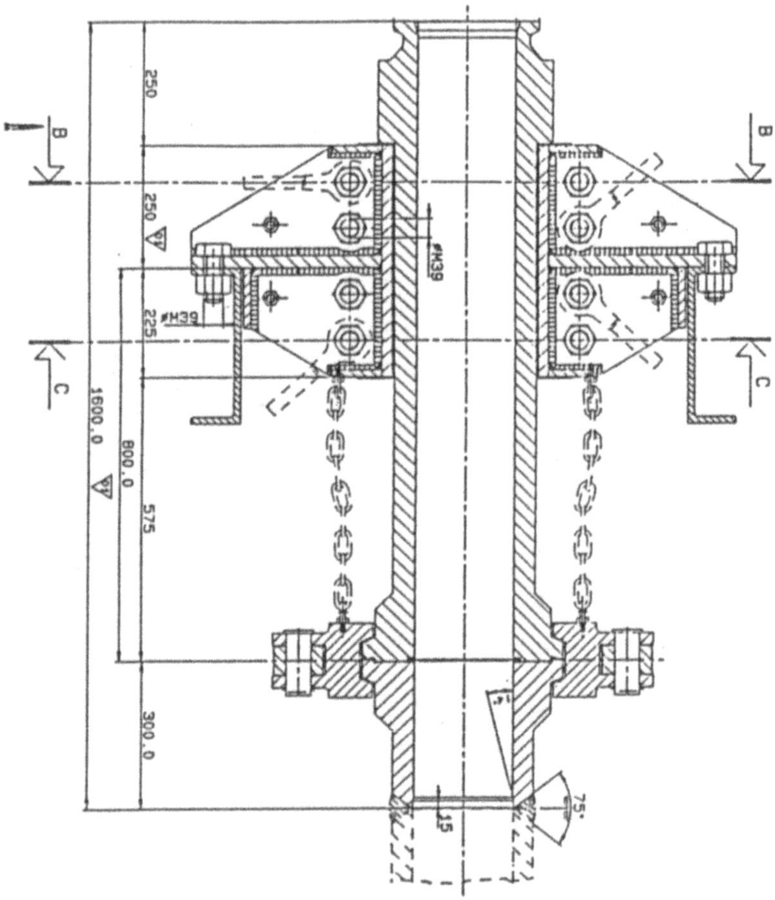

FIG. 7: Emergency Release Type Mechanical Connector

CONCLUSIONS

The development of flexible pipe began in the 1970s as a straightforward attempt to find an alternative pipeline design which could by its inherent flexibility be readily transported, quickly installed, and which could accomodate ever greater misalignments. From its early application as seabed spool pieces or flowlines a proliferation in product range and, consequently, applications has developed. This development has evolved to a point where today´s multi-riser systems can adapt numerous seabed well arrays to a variety of different floating production vessels.

Early pipe designs were previously regarded with an air of mystery and the designer´s art was known to be similar to a "black box technology". The introduction however of new manufacturers and products, such as the "Bonded" type pipe in the 1980s has made this industry more open and competitive as well as making the oil industry more keen to unravel hidden pipe mysteries. Today a better understanding of this new technology has been gained through industry initiatives largely concerned with improving knowledge of the composition and interrelationship of pipe layers. Correlation of measured data with computer predictions of stress levels has been greatly improved. Advances in computer software and enhanced processing facilities together with data gained from long term dynamic mechanical test results today permit more accurate determination of pipe performance. Advances in pipe and ancillary hardware design permit flexible pipe to play its role as a "key" mechanical subsystem in future subsea and floating production systems.

REFERENCES

1. American Petroleum Institute (1987). API RP17b-Recommended Practice for Flexible Pipe, API, Houston.

2. Boef, W., Lange, F., Van den Boom, H. (1989). Analysis of Flexible Riser Systems, Floating Production Systems, Conference procs. 11/12 Dec. ´89, IBC, London.

3. Kokkinowrachos, K., Giese, K. Peuker, M. (1987) Development of flexible risers for floating Offshore production. OTC 5469, Houston.

4. Mckeehan, D., Halkyard, J., Griffiths, R. (1986). Marine Applications for a Continuous Fibre Optical Strain Monitoring System, OTC Procs. 5119, Houston.

5. Neffgen, J.M. (1988). Integrity Monitoring for Flexible Pipes, from Pipes and Pipelines International, May/June, London.

6. Neffgen, J.M. (1989). Optimisation of Subsea and Floating Production Systems Utilising Flexible Pipe Technology, from Procs. Floating Production Systems, IBC, Dec. 11/12, London.

7. Often, O., Loetveit, S.A. (1989). Increased Reliability through a Unified Analysis Tool for Bonded and Unbonded Flexible Pipes. In Advances in Subsea and Pipeline Technology and Engineering, Graham & Trotman, London.

8. PAG-O-FLEX (1987). Fatigue Testing of Flexible Risers. Joint Industry Project Report, Pag-O-Flex, Düsseldorf, W. Germany.

9. Ross, R.B. (1989). Investigation of Failures. Procs. Offshore Europe, SPE 19273, Aberdeen.

10. Sterzenbach, M. (1989). SOLS for Floating Production Systems. Procs. Floating Production Systems, 11/12 Dec. 89, London.

11. Veritec (1987). Guidelines for the Design and Construction of Flexible Pipes. Joint Industry Report JIP GFP-02, Oslo.

12. Wallen, B., Abrahamsen, T. A Stainless Steel for North Sea Applications. Procs. Norges Ingeniørs Organisation Conference, 30 Jan. 1984, Oslo.

DESIGN AND MATERIALS CONSIDERATIONS FOR HIGH PRESSURE FLEXIBLE FLOWLINES

Colquhoun, R.S., Hill, R.T., and Nielsen, R.

Wellstream Corporation, 800 Bering Drive, Houston, Texas

Abstract

With advances in marginal field development, subsea completions, and production step-out programmes, the need for small diameter high pressure flowlines has continued to increase. To meet this demand a wide spectrum of project-specific unbonded layered flexible pipes have been engineered using state-of-the-art computer-aided design and materials selection technology. This paper highlights the circumstances in which such products offer technical and economic advantages over traditional steel pipe solutions.

Volume 24: C. P. Ellinas (ed.), Advances in Subsea Pipeline Engineering and Technology, 145–178.

INTRODUCTION

The quayside cost of flexible pipe pipe has hitherto generally been several times the quayside cost of corresponding rigid steel pipe. The use of flexible pipe has therefore been confined to applications where there was an obvious inherent technical advantage, for example dynamic risers and subsea jumpers, or where there was an economic advantage associated with the shorter mobilization time, lower dayrate, and faster laying speed of the flexible pipe deployment vessel - it is the installed cost on the seabed that is of interest to the operator.

The scope for using flexible pipe is widening rapidly, however, as a result of the increasing number of subsea well tie-backs involving aggressive process fluids and difficult environmental conditions and as a result of the increasing importance of extending the installation season in the interests of achieving early production. The expanding demand is being met by advances in materials technology and a re-thinking of the traditional criteria for the use of flexibles. Competitive flexibles for unsophisticated applications are also now available.

This paper does not specifically address dynamic risers or subsea jumpers (though many of the considerations addressed apply equally in those cases) since the need for flexibles in those applications is usually obvious. Attention is centred rather on subsea flowlines and on the opportunities which taylored flexibles have to offer in this context.

Reference is made to authors who have previously addressed the uses of flexible pipe. Among these are Dubois 1980, Strader 1982, Colquhoun 1982, and Nielsen 1983.

SUMMARY OF FLEXIBLE PIPE INDICATORS

Flexible pipe is likely to offer technical and economic advantages over rigid steel pipe in the following situations:

* Isolated flowlines of limited length where installation mobilization costs are predominant, Colquhoun 1988;

* Flowlines of all lengths in situations where the need for early production necessitates short weather-window off-season installation;

* Flowlines in extended well-test situations or other limited operational life situations which favour the flexible, which can be recovered and redeployed;

* Flowlines in deep water, where the flexible is much more easily and quickly installed than the rigid pipe, Strader & Nielsen 1982;

* Flowlines of all lengths with sweet low-pressure process conditions ("sweet & low") which enable cheaper materials to be used;

* Flowlines of all lengths in which the aggressiveness of the process fluid would require a rigid pipe to be made of expensive stainless steel or would require a significant capital and operational expenditure on a corrosion inhibition system, Hill 1989 (OTC 6114) and Hill 1989 (Corrosion 89);

* Flowlines of all lengths in which the production fluid leaving the wellhead is at an elevated temperature (typically above 80°C) which would result in wax dropout or hydrate formation due to falling temperature along the line in the absence of thermal insulation.

Each of these indicators is studied in greater detail below, but in order to give a better understanding of some of the flexible product structural design and materials considerations involved, reference should be made to the presentation of the concept of unbonded flexible pipe in Appendix A and the presentation of typical materials used in Appendix B.

It is difficult to generalize about the costs of flexible pipes, since there are so many possible product design variants and because, as for any batch production process, the length of product of given design produced at one time has a profound influence on production cost. As a general indication, to be treated with due circumspection, the cost of the raw materials may represent 25% to 50% of the ex works price of the pipe; the cost of production may represent from 1/100th to 1/1500th of the factory dayrate (capex charges plus O&M) per metre run of pipe depending on the run length and on the size and sophistication of the product in question. The amount and method of application of the OH&P depends on commercial policy and varies from one manufacturer to another.

REVIEW OF FLEXIBLE PIPE INDICATORS

Isolated Short Lines

The as-laid cost per unit length of flowline may be calculated as:

$$C_L \doteq C_Q + D_1{*}N/L + D_2{*}T/L + D_2/R$$

in which:

- C_Q is the cost per unit length of the pipe delivered to the operational shorebase;
- D_1 is the mobilization dayrate;
- D_2 is the operational dayrate of the lay vessel or marine lay spread including bunkers;
- R is the laying speed (length per day);
- N is the number of days required for mobilization, demobilization and transit to & from site;
- L is the total length of line laid in association with this mobilization;
- T is the total transiting time to/from the worksite (days).

In situations of moderate pressure (207bar/3000psi) and moderate temperature (60 to 80°C) which do not involve highly corrosive process fluids or thermal insulation requirements, the value of C_Q will typically be 3 to 5 times higher for flexible pipe than for coated rigid steel pipe. On the other hand the installation spread dayrate for flexibles is typically half of that for welded rigid steel pipe, the mobilization time for a flexible lay vessel is typically a half of that for a rigid pipe lay spread, and the laying speed with the flexible may be as much as 1 knot, ie: 50 km/day as against say 3000 metres/day for rigid pipe.

If typical historical cost data are inserted in the formula, the break-point for the economical use of flexibles under these conditions will be found to lie in the region of 1 to 5 kilometres. Nevertheless recent experience with medium depth lines in the Gulf of Mexico has demonstrated that flexibles can be more economical than rigid pipe for lengths up to at least 20 km.

Installation considerations in connection with the flowline terminations, both at the wellhead and at the platform, are one of the factors which can swing the balance heavily in favour of flexibles. Rigid risers are awkward to handle; flexible (static) risers are much simpler to deal with. Moreover there are problems and costs associated with the seabed expansion devices needed to protect rigid pipes and connected structures from each other when temperature changes would otherwise induce large deflections and reaction forces; these problems are absent in the case of flexible flowlines.

Temperature-induced upheaval buckling is less of a problem with flexibles than with rigid lines. On the other hand flexibles are susceptible to pressure-induced upheaval buckling in the absence of an adequate awareness of the phenomenon and of the techniques available for avoiding it. These include trenching under pressure as well as the sand/rock dumping techniques pioneered by Westminster-Seaway and now well-established in the North Sea.

Early Production Situations

In many parts of the world the installation season is limited. For example in the North Sea, North Atlantic, and European arctic waters the installation season for rigid steel pipe is generally confined to the 6-

month summer period from April to September. Attempts to lay rigid pipe during the winter half-year are normally associated with a disproportionately high risk - indeed a certainty - of substantial weather downtime because of the very short fine-weather windows during these months.

A flexible pipe lay vessel on the other hand is able, because of its high transit speed and high laying speed, to make use of a single quite short weather window to "get in, get it over with, and get out". Moreover in the event that deteriorating weather makes abandonment and recovery (A&R) operation unavoidable, the attendant risks of pipe buckling are much less in the case of the flexible.

The economic advantage to the operator of bringing forward the execution of an installation operation which may contribute to advancing the production startup date by weeks or even months will generally be sufficient to make any increased capex associated with the choice of flexible pipe pale into insignificance. In these situations it will be appropriate to select flexible pipe even when the lengths of flowline involved are much greater than the theoretical break-point calculated on the basis of the simple considerations in the preceding section.

Extended Well-Test and Other Short Life Operations

Flexible pipes can be recovered from the seabed without great difficulty and can be redeployed at a new location. Rigid steel pipe cannot usually be recovered conveniently or economically. Redeployment of a flexible means that the capex can be written down over more than one operation. This fact has broader financial implications, since it enables the concept of leasing flexible pipe to be exploited. This kind of financial "new thinking" is becoming increasingly important in the realm of marginal field development.

An obvious example of a limited operational life situation is the extended well-test, where hydrocarbon production may be continued for a year or two after completion of the well-test by using the well-test configuration, thus providing the operator with a cash flow during the lean period during design and installation of permanent facilities.

Deep Water Flowlines

In water depths beyond 300 metres (1000 feet) the cable lengths required for station-keeping with anchors cannot be accommodated. For most anchor barges the operating depth limit is much less than this. In deep water full dynamic positioning is required. The number of rigid pipe lay barges with DP in the world is small, their dayrates are high, and their transit speeds are low. The single station joining techniques required for steep angle deepwater rigid pipe laying, whether by welding or by mechanical connection, are time-consuming and may be of questionable quality in the as-laid condition.

Furthermore, deep water situations are characterized by dynamic risers, and this means that vessels and equipment for deep water flexible pipe deployment will in any event have to be mobilized to the site. This tips the balance in favour of the selection of flexibles for the flowlines as well.

Mild Production Conditions

There are many fields, notably in the Gulf of Mexico but also in other locations including the North Sea, where corrosive conditions associated with the presence of H_2S and CO_2 in the produced fluids are mild, the pressure is around 100 to 200 bar (1500 to 3000 psi), and the process and ambient temperatures are well within the range -20°C to +60°C.

The absence of severe corrosivity enables carbon steel to be substituted for stainless steel in the carcass; this reduces the cost of the carcass material by a factor of 5.

Full-scale tests at the Danish Corrosion Centre have confirmed that plain carbon steel carcass is in fact viable in a much wider range of process fluid corrosivity conditions than was formerly believed.

Dawans 1986 (OTC 5231) reports that HDPE is quite suitable for long exposure to hydrocarbons below 60°C. Thus the moderate Gulf of Mexico pressure and temperature levels enable HDPE to be substituted for Nylon 11 in the barrier layer; this reduces the cost of the barrier layer material by a factor of almost 8. The outer seal and intermediate wear layers can likewise be of HDPE.

A cautionary note is important in relation to the selection of the correct type of HDPE for oil and gas industry service. Poor performance can be expected from any HDPE made by the Ziegler process, and poor performance can be expected from HDPEs that were made 20 years ago. Significant improvements in HDPE for oil and gas service have resulted from the introduction of the Phillips process some 10 years ago and from the process developed more recently by Chevron. HDPE made by the new Chevron process has two particularly important characteristics: (a) improved stress crack resistance through a catelyst improvement that leads to optimal molecular weight distribution (no stress crack failures in service have been reported for pipes made from this material); and (b) improved stabilizer packages to eliminate thermal degradation during processing, to prevent oxygen attack, and to extend UV protection of the finished pipe.

Chevron currently makes and sells polyethylene drain pipes which perform extremely well in both fresh and salt water service. Chevron gas transmission pipe is also often exposed to fresh or saline groundwater without deleterious effects. The Chevron polyethylene is completely impervious to attack by petroleum. Only exceptionally vigorous surfactants, of a kind which do not occur naturally in petroleum, can affect the material by leading to some swelling at temperatures above 60°C, which is thus the upper bound service temperature for HDPE flexibles.

The moderate pressure level also means that no zedform layer is required. Furthermore the absence of H_2S means that the helical armour can be of high tensile steel, which roughly halves the weight of steel required for this layer (unless the pipe is in shallow water where the extra weight is required for on-bottom stability) and substantially reduces material cost since the premium on high tensile steel is only about 40%.

The upshot of these materials substitutions is that flexible pipe can compete economically with rigid steel pipe for small diameter (ie: 4" and

6") lines regardless of length in water depths down to 40 metres or less when the production conditions are identified as mild.

Flowlines for Aggressive Production Fluids

When the production conditions are such that a rigid steel pipe made from standard API linepipe would be unacceptable and a stainless steel alloy would have to be substituted, the balance tips in favour of the flexible. This is partly because of basic material cost, whether solid or clad pipe is specified, and partly because of the greater complexity and slower output which characterise the execution of full-penetration butt welds in either solid or clad rigid pipe, Hill 1989 (2 references).

The key properties of low-carbon and stainless steels and their areas of application in relation to the corrosive characteristics of the process fluid are summarized in Appendix B. The material cost of 758 MPa 304L is some 5,5 times that of 345 MPa low-carbon steel; the jump to 316L adds some 5%, and a further jump to Duplex adds another 10%; selection of high-nickel stainless steel adds yet another 20%. It should be noted that the decisive criterion in relation to the aggressiveness of H_2S and CO_2 is not moles % concentration but partial pressure.

The provision of corrosion inhibition in the form of a continuous or batch system is sometimes considered as an alternative to the use of high alloy grades for rigid pipe. In this case the comparison between the rigid and flexible pipe alternatives should include not only the capex of the injection, recovery, and recycling system, but also the NPV of the operational costs of the system including recycling logistics and the makeup of chemical loss.

Flowlines Requiring Thermal Insulation

When the production fluid leaving the wellhead is a waxy crude or wet gas at a relatively high temperature, there is a danger of flowline blockage due to wax dropout or hydrate formation if the temperature at any point in the line is allowed to fall below a critical level which is governed by the composition of the flow and the temperature and pressure conditions. In these cases the flowline must be thermally insulated.

The insulation of rigid steel pipelines is expensive. It involves the addition of an external steel carrier pipe, the provision of support and compartmentalization diaphragms, and the injection of foamed plastic insulation material into the annulus. This work has to be carried out in a special shorebased facility, and the lengths of pipe taken to the offshore location have to be two orders of magnitude longer than the 12 metres of the standard uninsulated pipejoint. Altogether different installation techniques have to be employed, and the cost of insulated rigid steel pipe is several times that of the equivalent uninsulated rigid steel pipe.

A flexible pipe by contrast is relatively easy to insulate by wrapping a suitable kind of high voids-ratio tape around the intermediate product prior to application of the outer plastic seal. The adverse corollaries are modest: a slight increase of pipe diameter, and a corresponding reduction of the length of pipe that can be accommodated on a given reel. The percentage increase in the cost of the pipe is likewise modest.

CUSTOM-DESIGN

Unbonded flexible pipe is standard in its general structure, but is an infinitely variable product when it comes to detail. Changes can be made in the selection of materials and in the dimensioning of the individual layers and their components so as to provide a wide range of products with different physical (weight and buoyancy), mechanical (strength and stiffness), and chemical (corrosion resistance) characteristics meeting the equally widely nouanced requirements of different field situations. The manufacturing processes are common to all the variants, but cost-effective application of the underlying concepts calls for an efficient engineering tool for product design and optimization for each specific job.

One such tool, dubbed "Pipemaker", has been developed and used by the authors and their colleagues. It is an interactive program built within the environment of Macintosh Microsoft Excel 2.2 and encompassing a database of materials properties and costs as well as the design theory presented in simplified form in Appendix A. Such a tool is a sine qua non for economic and reliable flexible pipe design.

Examples of four typical design variants are presented in Appendix C.

CONCLUSIONS

The rapidly expanding use of subsea developments and stepouts has precipitated a new approach to flexible pipe materials selection and a renewed awareness of the very wide range of applications where flexible pipe can compete technically and economically with rigid steel pipe in the context of submarine flowlines.

APPENDIX A - CONCEPT OF UNBONDED FLEXIBLE PIPE

Each of the layers of an unbonded flexible pipe is free to move relative to the others. That is to say, there is no composite action in bending or torsion. Each layer has its own specific function to perform. A typical structure consists of:

The Carcass Layer

The purpose of the carcass is twofold. The first is to provide support to the fluid barrier layer against the collapsing influence of external hydrostatic pressure. The second is to provide support against the squeezing action of the helical armour wire layers under the influence of axial tension; the effect on the carcass is equivalent to an external hydrostatic pressure. A primary constraint on the choice of format for this layer is the need for flexibility.

The carcass layer consists of an interlocked helix of steel strip pre-formed into an S-section. The resulting cylindrical structure is dimensioned to ensure Euler stability under the greater of the applied external pressure loads.

The critical pressure for collapse P_{cr} is calculated from the classical formulae for Euler stability of a solid cylinder with due allowance for out-of-roundness, but with the yield stress modified by multiplying by the "fraction filled" Ff, ie: the ratio between the solid cross-section area of the interlocked S-profile of the formed strip and the section area of the same sagittal length of the full layer thickness (or expressed another way, the ratio of the average thickness of steel in the layer to the total thickness of the layer). The validity of this calculation technique has been established in tests.

The collapse pressure is thus obtained from the equation:

$$P_y^2 + B*P_y + C = 0$$

in which:

$B = -[(\sigma_y*Ff/m) + (1 + 6*m*n)(P_{cr})]$

$C = \sigma_y*Ff*P_{cr}/m$ $m = R/h$ $n = u_o/R$

σ_y = yield stress of carcass strip Ff = "fraction filled"

R = mean radius of carcass layer h = thickness of layer

u_o = initial ellipticity

$P_{cr} = (h/R)^3*(1/4)[E/(1-\mu^2)]$

E = flexural modulus of the carcass strip

μ = Poisson's ratio for the carcass strip

The carcass strip may be of plain carbon steel or of any one of several grades of stainless steel depending on the process fluid conditions to be encountered.

Under conditions of extreme corrosivity the carcass layer may be omitted and its role performed by the plastic barrier layer in conjunction with the steel zedform layer, McCone 1989.

The Barrier Layer

The purpose of the plastic barrier layer is to contain the production fluid. It is restrained against collapse by the carcass beneath and against burst by the armour layers above.

The thickness of the barrier layer must be such that the plastic will not yield under the influence of internal pressure where it spans the gap between two armour wires. When there is no circumferential pressure armour this gap is equal to: $\pi \times D \times (1 - Ff) \times \cos.\phi /N$ in which D is the diameter at the layer interface, Ff is the helical armour layer fraction-filled, ø is the helical armour lay angle, and N is the number of armour wires. There must be some allowance for manufacturing tolerance.

The stress in the plastic can be obtained from the equation:

$$\sigma = K*(q*b^2)/(2*t^2)$$

in which:

q = the process fluid pressure

b = armour wire gap width

t = thickness of the plastic barrier layer

K = a multiplier reflecting deep beam stress concentration

For b/t	=		2,0	1,5	1,0	0,5	0,33
K	=	approx.	1,1	1,4	2,7	11	25

For a given thickness of the barrier layer the critical gap width falls off fairly rapidly with increasing pressure as a result of the effect of the stress concentration multiplier.

When there is an interlocked circumferential pressure (zedform) armour, the gap width is very small, and the required barrier layer thickness is limited to that which ensures that age-hardening cracks will not penetrate the wall during the lifetime of the pipe. For practical manufacturing reasons the barrier layer thickness is usually not less than 5mm.

An additional factor in dimensioning the barrier layer is the need for controlling the flow rate of hydrocarbon and other gas diffusing through the layer. This is a straightforward calculation in accordance with the equation:

$$Q = c*p*S/t$$

in which:

Q is the volume flow of gas in cc/sec

c is the permeability coefficient in cm^3-cm/sec-cm^2-bar

(See Fig.B.2 after Dawans et al.)

p is the gas pressure in bar

S is the diffusion surface area in cm^2

t is the thickness of the thermoplastic

but the calculation must be undertaken within the broader context of the overall gas diffusion and annular gas flow regime of the system. In calculating the diffusion flow through the barrier layer it is important to note that this flow is confined to the gap area between the armour wires, since the area covered by the armour wires represents a path whose resistance is several orders of magnitude higher. This does not mean that the diffusion flow can simply be reduced in proportion to (1 - Ff). The flow through the plastic is 3-dimensional, and the factoring must be based on a more nouanced computation. It is also important to note that the coefficient of permeability of thermoplastics is heavily temperature-dependent, and since the process fluid temperature is generally well above ambient seawater temperature, this means that gas diffusion through the outer plastic shield protecting the armour will be negligible compared with the gas diffusion through the plastic barrier layer. It is therefore important to ensure that the diffusion gas can escape from the annulus in a satisfactory manner either via burst discs in the outer shield or via flow along the armour wire gaps in the annulus to vents at the end fittings.

In cases where the carcass is omitted because of highly corrosive fluid conditions (eg: acid flow in service lines), the barrier layer may nonetheless be stable under the influence of nett external hydrostatic pressure provided that there is a zedform layer around it to restrain it against Euler collapse due to "rosetting". The critical pressure is given by:

$$p_{cr} = [E/(1-\mu^2)][(h/R)^3/12](n^2-1)$$

in which:

E	=	elastic modulus of the plastic
μ	=	Poisson's ratio of the plastic
h	=	thickness of the barrier layer
R	=	radius of the barrier layer
n	=	number of deformation wave periods around the circumference; n must be an even number; when the barrier layer is unrestrained (ie: there is no zedform layer), n = 2.

The actual value of p_{cr} is obtained from the equation using the largest value of n compatible with stability. The largest value of n for stability is given by:

$$n^2 < R/v$$

where v is the gap between the plastic barrier layer and the steel zedform layer after elastic compression of the plastic under the influence of the nett external pressure. The value of v is obtained from:

$$v = R*p*D/2h*E$$

in which p is the hydrostatic pressure.

The plastic most widely used hitherto for the barrier layer has been Nylon 11, which combines strength, suppleness, low water absorbtion, resistance to hydrocarbon attack, and slow age-hardening at temperatures up to approx. 90/110°C. For moderate pressure non-corrosive applications up to 60°C HDPE (high density polyethylene) is a significantly cheaper alternative. PVDF (Atochem Foraflon) is available for service in the temperature range up to 130°C but is typically 60% more expensive than Nylon 11. A diagram can be drawn up relating physical degradation to duration of exposure at different temperatures as shown in Fig.B.1 after Dawans et al. 1986. This can be used to determine cumulative ageing degradation under the influence of a temperature regime which varies during the operation life.

Circumferential Pressure Armour (Zedform) Layer

The purpose of this layer is to support the plastic barrier layer against burst under the influence of the process fluid pressure within the pipe. In low pressure pipe this function can be performed satisfactorily by the helical armour layers, but in high pressure pipe the need to ensure that the width of unsupported plastic at the armour wire gaps is small dictates the use of an interlocked zedform circumferential armour layer immediately outside the plastic barrier layer.

Since there is always a double layer of contra-wound helical armour regardless of the presence of the zedform armour layer, the zedform layer has only to contribute a proportion of the total burst resistance of the pipe.

The burst pressure value of the zedform layer can be obtained from a modification of the Barlow formula as follows:

$$p_z = 2t * \sigma_y * Ff/D$$

in which:

σ_y = the yield stress of the zedform wire

D = the mean diameter of the zedform layer

t = the thickness of the zedform layer

Ff = the "fraction filled" of the zedform layer

For manufacturing reasons the zedform wire is of low-carbon steel with a typical yield stress of 758 MPa.

When the burst pressure requirement is high, the zedform layer may be supplemented by a second layer of circumferential armour. As this is not in direct contact with the plastic barrier layer, it does not have to be interlocked and can therefore be made of simple rectangular wire with similar chemical and mechanical properties.

Helical Armour Layers

There are always two contra-wound layers of helical armour wire. The wire is flat. It is pre-formed in torsion and bending so as to "kill" it on the pipe, ie: minimize the residual elastic stresses and prevent "birdcaging" when the wires are cut.

The purpose of the helical armour layers is twofold. The first is to support the plastic barrier layer against burst under internal process fluid pressure. When there is a zedform layer, this is a shared role. The second is to provide the pipe with axial tensile strength to enable it to resist the endcap force effects of internal pressure and to resist other tensile loads applied to the pipe, McCone 1989.

The helical armour contribution to burst pressure resistance is given by:

$$ph = 2t*\sigma_y*Ff*\sin^2\phi/D$$

in which:

t = total thickness of the double helical tensile armour

σ_y = yield stress of the armour wire

Ff = "fraction filled" in the armour wire layer

ϕ = lay angle of the armour wire, ie: the angle between the wire and the pipe axis

D = mean diameter of the helical armour layer

The helical armour contribution to endcap pressure resistance is:

$$p_a = 4t*\sigma_y*Ff*\cos^2\phi/D$$

In the absence of a zedform armour layer the helical armour alone resists both the hoop and the endcap effects of the internal pressure so that $p_a = p_h$. It can be seen that under these circumstances $\tan^2\phi = 2$, whence $\phi = 54{,}7^o$. This is the "balanced" lay angle at which there is no tendency for the armour helix to change shape under load.

When there is a zedform layer, the relationship changes and the "balanced" angle depends on the proportions of steel in the zedform and helical layers. The optimal lay angle is then typically 35° to 45°, but may lie outside this range in some cases.

When axial tension is applied to the pipe, the helical armour helix angle tends to decrease, but this is prevented by the support of the steel carcass beneath the barrier layer. The external pressure on the carcass is given by:

$$p_t = T*\tan^2\phi/(\pi*D^2)$$

in which T is the applied axial load. The stress in the tensile armour is:

$$\sigma = T/(2t*Ff*\pi D*\cos^2\phi)$$

The failure tension of the pipe may be governed by the collapse of the carcass or by the yield of the helical armour depending on the characteristics of these two layers.

The helical armour wires typically have cross-sections from 5x2mm to 14x3mm. The smaller sections are easier to handle and apply, but pipes of larger diameter require more armour, and the number of wires per layer is limited by the number of bobbins the armouring machine can accommodate (typically 72), so that larger pipes have to be armoured with wires of greater section in order to keep within the capability of the machine. The wires may be of low-carbon steel (say 758 MPa) or high-carbon steel (say 1550 MPa). The former is indicated when the process conditions are sour, since hydrogen sulphide and water vapour readily diffuse into the armour annulus. The low strength steel is also cheaper, and may therefore be preferable when the quantity of armour has to be increased to provide weight for on-bottom stability of the pipe.

De-Bonding Tape Layer

A thin layer of tape is wrapped around the outer helical armour layer in order to provide a bed for the outer shield plastic extrusion, ie: to prevent penetration of the hot extrudate into the armour interstices and thereby ensure that the outer seal layer is not bonded to the armour and that the armour layer lubricant is not expelled.

Outer Shield Layer

The purpose of the outer thermoplastic layer is to protect the armour layers from seawater and from mechanical damage. If both the barrier layer and the outer shield layer are of the same plastic, it is the OD of the latter which will normally govern the minimum bending radius of the pipe. The critical parameter is the maximum strain which the plastic can tolerate at the lower end of the operating temperature range without cracking. For Nylon 11, for example, it is 7,5% at -50°C. The minimum bending radius is given by: $R = D/2e$, in which D is the outside diameter of the outer plastic layer and e is the maximum acceptable strain. If the pipe is never to be exposed to an arctic atmosphere, then it may be possible to relax the bending radius limit to the point where other factors come into play, for example the geometric limitation inherent in the carcass and zedform armour layer construction or the risk of the helical armour wires "bunching up" and riding onto each other.

The bending stiffness of the pipe comes almost entirely from the plastic layers and is obtained from the expression $E(I_1 + I_2)$ in which I_1 and I_2 are the area moments of inertia of the barrier and outer seal layers. (Plastic wear layers, if present, will also contribute to the stiffness).

The outer shield layer may contain "burst discs", ie: small circular areas of wall thickness reduction designed to "pop" in the event of a buildup of diffusion gas pressure in the armour annulus.

Chevron HDPE is not attacked by either fresh or salt water, Janson 1974 p.47. Coflexip and IFP report (OTC 5231) that their tests on Nylon 11 likewise showed that this material was impervious to attack by seawater. Nylon 11 is more flexible and tough than the stiffer HDPE, and this characteristic is highly desirable in dynamic risers, but the 8:1 cost ratio relative to HDPE means that the use of Nylon 11 for the outer shield is not cost-effective in static applications unless unusual conditions exist.

Insulating Layer

When additional thermal insulation is required, this is provided in an extra layer between the helical armour and the outer plastic shield. One example is foamed PVC tape wound onto the pipe prior to application of the final thermoplastic extrusion. The thermal exchange calculations are similar to those familiar for insulated pipes generally, but all the component layers of the flexible pipe are taken into consideration, McCone 1989.

Wear Layers

Additional thin thermoplastic layers may be included, for example between the helical armour layers in order to stabilize the distribution of the armour wires during manufacture and handling, or outside the outer helical armour layer in order to separate an insulation layer from the armour.

APPENDIX B - PROPERTIES OF TYPICAL MATERIALS USED

Carcass Layer

Material	UTS (MPa)	Yield (MPa)	Elastic Modulus (MPa)	Fraction Filled	Application
Carbon steel	345	276	$2,1x10^5$	0,56-0,6	Sweet process
Stainless 304L	758	552	$2,0x10^5$	0,56-0,6	T<65C;NaCl<1%;CO_2<50b;H_2S<0,5b
Stainless 316L	758	552	$2,0x10^5$	0,56-0,6	T<65C;NaCl<5%;CO_2<50b;H_2S<0,5b
Duplex SS	758	552	$2,0x10^5$	0,56-0,6	T<150C;NaCl<10%;CO_2<40b;H_2S<1,4b
High Ni alloy	758	552	$2,0x10^5$	0,56-0,6	T>150C;NaCl>10%;CO_2>40b;H_2Sunlim

Thermal conductivity is 69 W/mK in all cases.

Barrier Layer

Material	UTS (MPa)	Yield (MPa)	Elastic Modulus (MPa)	Fraction Filled	Thermal Conductivity (W/mK)	Application Upper bounds are for 10 yr life
HDPE	34	21	758	1,0	0,3	T -20/+ 60°C
Nylon 11	55	29	345	1,0	0,3	T -40/+100°C
PVDF #	41	29	758	1,0	0,16	T -40/+130°C

Atochem Foraflon

Zedform Armour

Material	UTS (MPa)	Yield (MPa)	Elastic Modulus (MPa)	Fraction Filled	Application
Carbon steel	345	276	$2,1x10^5$	0,83	High pressure pipe

Thermal conductivity is 69 W/mK.

Helical Armour

Material	UTS (MPa)	Yield (MPa)	Elastic Modulus (MPa)	Fraction Filled	Application
Low-C steel	758	552	$2{,}1 \times 10^5$	0,9	Sour service
High-C steel	1550	1380	$2{,}1 \times 10^5$	0,9	Non-sour service

Thermal conductivity is 69 W/mK in both cases (less allowance for Ff)

De-bonding Tape

Material	UTS (MPa)	Yield (MPa)	Elastic Modulus (MPa)	Ff	Thermal Conductivity (W/mK)
Nylon fabric in double wrap totalling 0,3mm thick	55	29	345	1,0	0,3
Mylar in double wrap totalling 0,3mm thick	55	29	345	1,0	0,3

Insulating Layer

Material	UTS (MPa)	Yield (MPa)	Elastic Modulus (MPa)	Fraction Filled	Thermal Conductivity (W/mK)
Foamed PVC tape in 5mm thick wraps	20	14	620	1,0	0,06
Polypropylene webbing in 1,5mm thick wraps	20	14	620	1,0	0,08

Outer Shield and Wear Layers

The outer shield and wear layers are typically made of HDPE unless the temperature regime calls for a plastic with a wider temperature range rating or a reduced bending stiffness is required.

Fig.B.1 (After Dawans, OTC 5231)

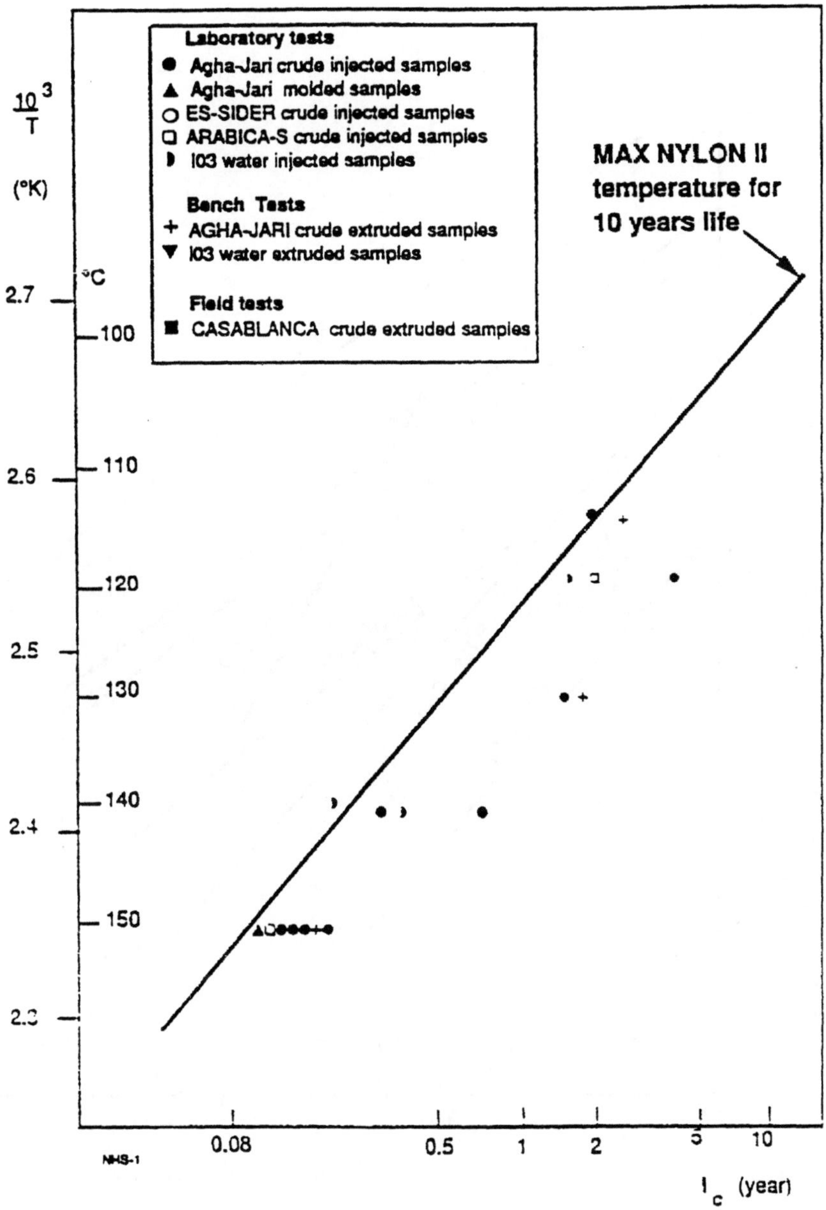

Critical time vs temperature curve for Nylon II

Fig.B.2 (Nylon 11 after Dawans, OTC 5231)
(Foraflon data from Atochem Inc.)
(HDPE data from Chevron)

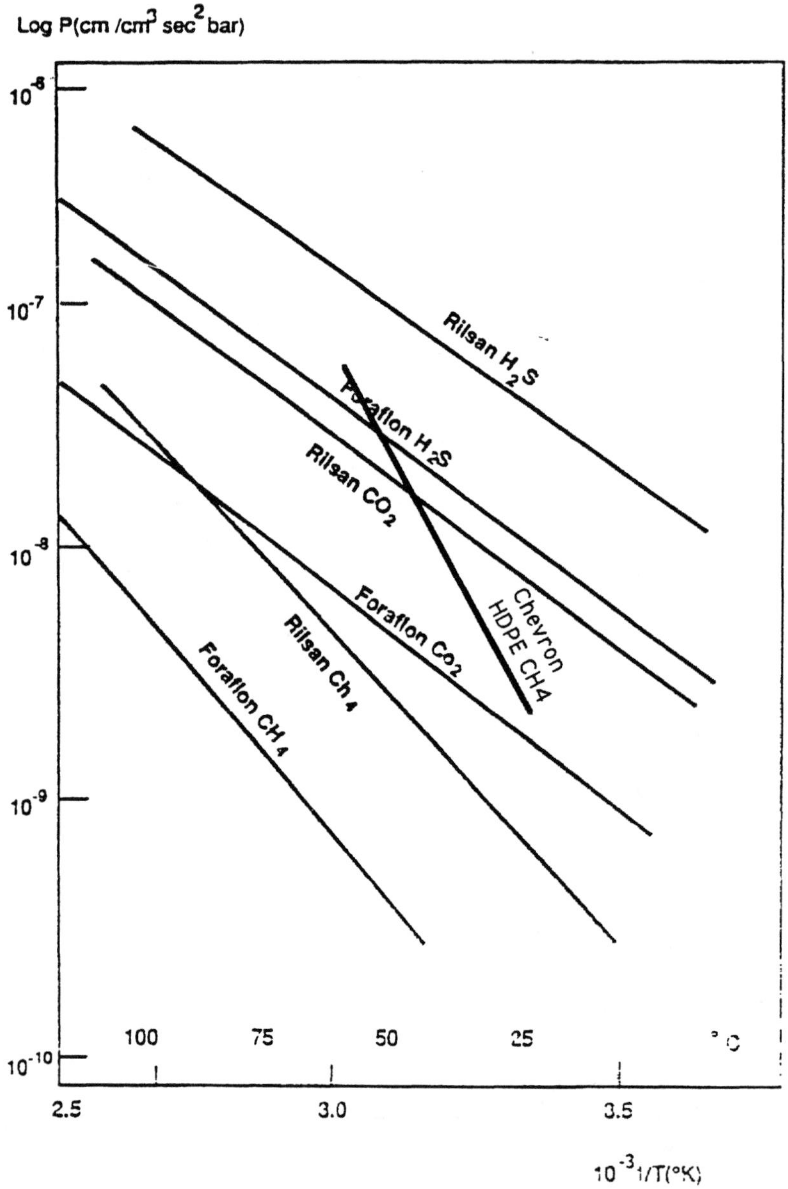

Permeability of Rilsan and Foraflon to H_2S, CO_2 and CH_4 Gases

APPENDIX C - STRUCTURE AND PROPERTIES OF SOME TYPICAL DESIGN VARIANTS

(1) 4" ID 100 Bar (1450 psi) "Sweet and Low" Flowline

Layer	Description	Thickness (mm)	Weight (kg/m)
Carcass	Low-carbon at 276 MPa yield	5.0	7.9
Barrier	HDPE at 21 MPa yield	5.0	1.7
Double helical	High-carbon at 1380 MPa yield	4.0	11.7
Intermed. wear	HDPE at 21 MPa yield	2.0	0.5
Tape	Mylar	0.3	0.1
Outer shield	HDPE at 21 MPa yield	5.0	1.4

Property	English		Metric	
Inside diameter	4	inches	101.6	mm
Outsidse diameter	5.68	inches	144.2	mm
Weight empty in air	16.3	lb/ft	24.3	kg/m
Seawater content	5.6	lb/ft	8.4	kg/m
Seawater buoyancy	11.2	lb/ft	16.7	kg/m
Wt. seawater filled in air	21.9	lb/ft	32.7	kg/m
Wt. empty submerged	5.1	lb/ft	7.6	kg/m
Wt. seawater filled submerged	10.7	lb/ft	16.0	kg/m
Burst pressure	4283	psi	295	bar
Collapse pressure	1231	psi	85	bar
Failure tension	47,923	lb	21.7	tons
Minimum storage bending radius	3.2	ft	1.0	m
Elongation at design pressure	0.6	%	0.6	%
Elongation at 50 kN tension	0.25	%	0.25	%
Thermal exchange coefficient	2.5	Btu/hr-ftF	4.4	W/mK

(2) 6" ID 345 Bar (5000 psi) 90ºC and 30 Bar CO_2 Partial Pressure Flowline (Non-Sour)

Layer	Description	Thickness (mm)	Weight (kg/m)
Carcass	Stainless 304L at 550 MPa yield	6.0	13.1
Barrier	Nylon 11 at 29 MPa yield	7.0	4.0
Zedform	Low-carbon at 550 MPa yield	6.2	23.7
Double helical	High-carbon at 1380 MPa yield	6.0	26.5
Tape	Neoprene	0.3	0.1
Insulation	Foamed PVC tape at 0.06 W/mK	7.6	2.7
Outer shield	Nylon 11 at 29 MPa yield	7.0	4.9

Property	English		Metric	
Inside diameter	6	inches	152.4	mm
Outsidse diameter	8.56	inches	217.4	mm
Weight empty in air	48.5	lb/ft	72.2	kg/m
Seawater content	12.5	lb/ft	18.7	kg/m
Seawater buoyancy	21.8	lb/ft	32.5	kg/m
Wt. seawater filled in air	61.0	lb/ft	90.8	kg/m
Wt. empty submerged	23.0	lb/ft	34.2	kg/m
Wt. seawater filled submerged	35.5	lb/ft	52.8	kg/m
Burst pressure	12,221	psi	843	bar
Collapse pressure	975	psi	67	bar
Failure tension	579,273	lb	258	tons
Minimum storage bending radius	4.0	ft	1.2	m
Elongation at design pressure	0.9	%	0.9	%
Elongation at 50 kN tension	0.04	%	0.04	%
Thermal exchange coefficient	6.9	Btu/hr-ftF	12.0	W/mK

(3) 3.5" ID 300 Bar (4350 psi) Multi-Purpose Service Line (Non-Sour)

Layer	Description	Thickness (mm)	Weight (kg/m)
Barrier	Nylon 11 at 29 MPa yield	5.0	1.5
Zedform	Low-carbon at 550 MPa yield	6.2	13.5
Double helical	High-carbon at 1380 MPa yield	4.0	10.2
Tape	Neoprene	0.3	0.1
Outer shield	Nylon 11 at 29 MPa yield	5.0	2.1

Property	English		Metric	
Inside diameter	3.5	inches	88.9	mm
Outsidse diameter	5.12	inches	130.0	mm
Weight empty in air	18.4	lb/ft	27.4	kg/m
Seawater content	4.3	lb/ft	6.4	kg/m
Seawater buoyancy	9.1	lb/ft	13.6	kg/m
Wt. seawater filled in air	22.7	lb/ft	33.8	kg/m
Wt. empty submerged	9.3	lb/ft	13.8	kg/m
Wt. seawater filled submerged	13.5	lb/ft	20.1	kg/m
Burst pressure	14,500	psi	1000	bar
Collapse pressure	1,000	psi	69	bar
Failure tension	297,100	lb	132	tons
Minimum storage bending radius	2.1	ft	0.64	m
Elongation at design pressure	0.34	%	0.34	%
Elongation at 50 kN tension	0.04	%	0.04	%
Thermal exchange coefficient	4.0	Btu/hr-ftF	6.9	W/mK

(4) 8" ID 300 Bar (4350 psi) Flowline Insulated to 3.5 W/mK, Wellhead Temperature 90ºC, Design Life 20 Years Possibility for Moderately Sour Conditions

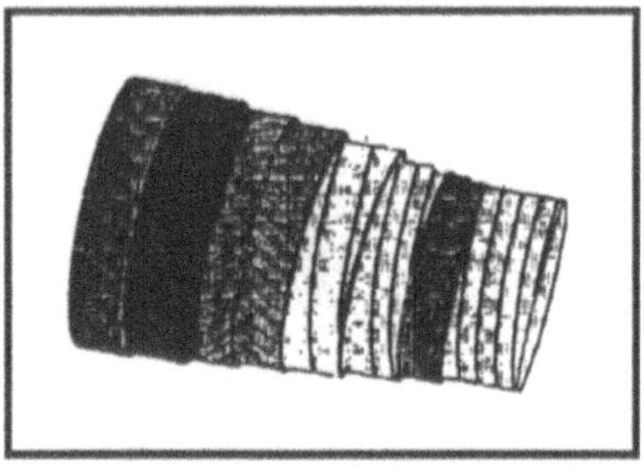

Layer	Description	Thickness (mm)	Weight (kg/m)
Carcass	Stainless 304L at 550 MPa yield	7.0	20.7
Barrier	Nylon 11 at 29 MPa yield	7.0	5.2
Zedform	Low-carbon at 550 MPa yield	6.2	30.5
Flat circ. strap	Low-carbon at 550 MPa yield	6.0	34.2
Double helical	Low-carbon at 550 MPa yield	6.0	35.6
Tape	Neoprene	0.3	0.3
Insulation	Foamed PVC tape at 0.06 W/mK	7.6	2.7
Outer shield	Nylon 11 at 29 MPa yield	7.0	6.7

Property	English		Metric	
Inside diameter	8	inches	203.2	mm
Outsidse diameter	11.71	inches	297.5	mm
Weight empty in air	91.1	lb/ft	135.6	kg/m
Seawater content	22.3	lb/ft	33.2	kg/m
Seawater buoyancy	47.8	lb/ft	71.2	kg/m
Wt. seawater filled in air	113.4	lb/ft	168.8	kg/m
Wt. empty submerged	43.3	lb/ft	64.4	kg/m
Wt. seawater filled submerged	65.6	lb/ft	97.6	kg/m
Burst pressure	9677	psi	667	bar
Collapse pressure	722	psi	50	bar
Failure tension	740,483	lb	336	tons
Minimum storage bending radius	5.5	ft	1.7	m
Elongation at design pressure	0.17	%	0.17	%
Elongation at 50 kN tension	0.006	%	0.006	%
Thermal exchange coefficient	2.0	Btu/hr-ftF	3.5	W/mK

ACKNOWLEDGEMENTS

The authors wish to express acknowledgement to their many colleagues at Wellstream Corporation for their contributions to the material included in this paper.

Acknowledgement is also due to flexible pipe plastic and steel suppliers in the U.S.A. and U.K. for information and advice on specific materials. Special thanks are due to Atochem Inc. for their support.

REFERENCES

Atochem Notice No. 694.E/07.87/20: "Foraflon-PVDF".

Colquhoun, R.S.: "Flowline As-Laid Costs - Comparison of Flexpipe and Standard API Linepipe", Copenhagen, 1988, Wellstream Corporation Technical Note Series.

Colquhoun, R.S., and Nielsen, R.: "Inherent Properties of Linepipe for Sour Oil & Gas Service", Copenhagen, 1983 (Offshore Oil & Gas Pipeline Technology European Seminar, Oyez).

Colquhoun, R.S., Abbott, M.B., and Bryndum, M.: "Deepwater Pipeline Lay Analysis", Stavanger, 1982 (Deep Water Pipeline Technology Conference, Oyez).

Danish Corrosion Centre: "Full-Scale Carcass Corrosion Test Program", for Wellstream Corporation, Copenhagen, Jan. 1990.

Dawans, F.A., Jarrin, J., Lefevre, T.O., and Pelisson, M.: "Improved Thermoplastic Materials for Offshore Flexible Pipes", Houston, 1986 (OTC 5231).

Dubois, B.C.: "New Possibilities Offered by Using Flexible Pipe for Production Application in the Gulf of Mexico", New Orleans, 1980 (ASME Petroleum Division, Energy Technology Conference).

Hill, R.T., and Barnes, C.B.: "Economic Analysis for High-Alloy Offshore Pipelines", Houston, 1989 (OTC 6114).

Hill, R.T., and Christensen, C.: "Full Scale Test of Line Pipe Steels for Sour Service", New Orleans, 1989 (Corrosion 89).

Janson, L-E.: "Plastic Pipe in Sanitary Engineering", New York, 1974 (Gränges Essem Plast).

Nielsen, R., Colquhoun, R.S., and Gibbs, R.C.: "Pipeline & Flowline Installations in Deeper Water", Shanghai, 1983 (Marintec Offshore China Conference).

McCone, A.I.: "Technical Notes on Unbonded Flexible Pipe Product Design Methods", San Francisco, August 1989, Wellstream Corporation Technical Note Series. (Calculation of carcass collapse; Stability of pipes without carcass; Calculation of burst pressure; Calculation of axial stiffness; Bending stiffness & minimum radius; Thermal exchange and thermal ageing calculations; Gas diffusion calculations.)

Oliveira, J.G.de, Goto, Y., and Okamoto, T.: "Theoretical and Methodological Approaches to Flexible Pipe Design and Application", Houston, 1985 (OTC 5021).

Roark, R.J., and Young, W.C.: "Formulas for Stress and Strain (5th Edition)", McGraw-Hill, 1976.

Strader, J.B., and Nielsen, R.: "Design Aspects of Subsea Flowlines in over 1500 Feet Water Depth", Basel, Switzerland, 1982 (European Seminar on Offshore Oil & Gas Pipeline Technology, Oyez).

THE PROBLEMS ASSOCIATED WITH NDT OF HIGH PRESSURE FLEXIBLE PIPES

Authors. G.T.Ashcombe and Dr. R.C.Kenison BP Engineering.

ABSTRACT.

The variety of materials used in the manufacture of high pressure flexible pipes provides, in the first instance, an extensive range of possible NDT methods. However their complex, composite construction imposes equally extensive constraints on the operability and potential effectiveness of many of these methods.

Additionally, little detailed information is available on actual failure modes of flexible pipes and data from examinations of failed pipes is largely confidential to individual operators.

Ageing of non-metallic components presents a particular problem since nearly all of the common tests for determination of material properties are of a destructive nature.

In addition to the complex construction further constraints on in-service inspection are imposed by the presence of bend restrictors, anode

Volume 24: C. P. Ellinas (ed.), Advances in Subsea Pipeline Engineering and Technology, 179–204.

bracelets, and buoyancy elements on risers, by the burial of flowlines, and by encasement in J-tubes.

This paper addresses some of the problems of inspection and monitoring flexible pipes and describes some of the methods which are currently being considered as potential solutions to them.

INTRODUCTION

There has been noticeably growing interest over the last 2 to 3 years within the offshore industry, in establishing methods for condition monitoring of flexible pipes.

With greater use, and potential re-use, of these products the need for adequate inspection techniques is increasing.

The in-service inspection of flexibles is complicated by the subsea environment in which they are used and even when access to the pipe is gained the multi-layer composite construction makes quantitative assessment of pipe condition a demanding objective.

THE NEED FOR INSPECTION

The design of flexible pipe is a subject on which information still remains, for commercial reasons, largely proprietary to the suppliers.

Inspection during manufacture and quality control of materials is invariably monitored by the purchaser's representative and this should ensure that pipes of the highest quality are manufactured. However, for the user, his confidence in the pipe's fitness for purpose can only stem from the supplier's track record of similar products in similar operating environments, and possibly from testing of small samples in simulated conditions.

Additionally, there exists the possibility that actual service conditions will differ from those for which the pipe was "designed". It is, for example, possible that actual operating temperatures are higher than predicted. Product composition, in terms of H_2S, CO_2 and water cut may vary significantly during field life and could exceed predicted values. External loadings imported by extreme environmental conditions may exceed design values. Each of these cases is obviously undesirable and furthermore the operator will not necessarily be aware of their occurrence.

There therefore exists a requirement to be able to determine the condition of a flexible pipe at intervals during its service life and to confirm that it is still fit for purpose and safe to operate. Operational and economic constraints will invariably dictate that this is done in-situ.

FAILURE MODES

We have touched on some of the possible causes of operation outside design conditions, but even under expected service conditions there will be a deterioration in the condition of the pipe.

This leads us to the controversial subject of failure modes.

One of the major problems when addressing the subject of NDT of flexibles is the definition of the modes of failure which we are looking for. With the limited availability of information on historical failures and their causes, one is led largely to making assumptions. Nonetheless it is widely accepted that one of the most critical scenarios is leakage at end fittings particularly for the top connection on risers where tensile and bending loads are highest.

External damage to the outer sheath can lead to seawater ingress and, unless cathodic protection is provided, subsequent corrosion of the tensile armours, this too being most critical for risers.

Internally the pipe will be exposed to a variety of fluids, probably at elevated temperatures, and possibly carrying solids particles, sand etc. Here it is the innermost layers of the pipe construction i.e. internal carcass and inner polymeric liner that will degrade as a result of abrasion and chemical attack respectively.

With the exception of pipes with a continuous steel liner (e.g. Pag-o-flex) there will also be gradual but continuous permeation of gas through the structure when used on gas or liquid hydrocarbon duties.

Damage to the intermediate layers within the pipe wall can be caused by:-

- External impact of sufficient magnitude as to penetrate the outer sheath and armour wires and cause permanent deformation to the intermediate layers.

- Minor flaws, e.g. wire cracks or small foreign bodies in polymer layers, introduced during manufacture but which are exacerbated as a result of service conditions.

These are just a selection of the causes of pipe deterioration, possibly leading to failure, which can be experienced.

In an attempt to prioritise the pipe wall constituents for NDT techniques the manufacturers have been asked to assess the contribution each layer makes towards the pipe integrity in Tension, Bending, Torsion and pressure containment.

It was to be expected that the layers providing for example, pressure containment contribute little or nothing to tensile strength and vice versa. However from the survey the following priorities have been attributed to the various layers.

TABLE 1 - RESULTS OF SURVEY

PRIORITY	UNBONDED PIPE	BONDED PIPE	
		Interlocked Carcass	Corrugated Liner
1.	Inner plastic sheath	Lining rubber	Corrugated liner
2.	Armour wires	Reinforcing matrix	Reinforcing matrix
3.	Interlocking wire	Primary ply matrix	Carbon steel spiral
4.	Interlocked carcass	External cover	Adhesion layers
5.	Pressure back-up	Interlocked carcass	Elastomer liner
6.	External sheath	-	Anti-extrusion layer
7.	-	-	External cover

CANDIDATE INSPECTION METHODS

BP and Department of Energy commissioned the Harwell NDT Centre to perform a study to identify the inspection techniques which could be considered for flexible pipes.

The resultant list of some 21 methods ranged in sophistication from external visual inspection through magnetic and ultrasonic methods, to radiography. By considering these particular examples it is possible to develop a philosophy for the NDT of flexible pipes.

Visual inspection is a qualitative technique which, with the aid of of ROV's and/or internal TV cameras represents an example of a FIRST PASS TECHNIQUE. It would be relatively inexpensive and would identify those areas where obvious damage had occurred.

Magnetic methods such as Eddy currents, would be termed Induced Flux methods where a flux field of some kind is induced into the pipe structure and the interaction between the know input field and the structure is monitored.

In their simplest form such techniques would be qualitative in as much that an anomaly on the output signal trace would indicate areas for further investigation. However, with equipment development, both in sensors and signal processing and with operating experience, it may ultimately be possible to obtain a quantitative measure of the anomaly.

These are typical of SECOND PASS METHODS. They would be quite quick in operation since the operator would be looking for anomalies and ignoring results which are within pre-determined bounds.

Finally, X-Ray and associated techniques are likely to be able to give the most detailed information on the condition of pipe components but will be extremely slow as well as localised in their output. These are THIRD PASS METHODS and are likely to be used once a pipe has been taken out of service as a result of the indications from first or second pass inspections.

We should now like to discuss some of physical methods which have been given further attention and the problems which are envisaged before their acceptance for field application.

First Pass Methods

These methods would be applied on a regular planned basis.

Included in this category are visual and dimensional, internal and external examinations. Their purpose is essentially to identify the presence of defects in the pipe which have resulted in some deformation of either the inner liner or the outer sheath. External visual inspection would be either by diver or ROV with underwater video cameras. Blistering or ruptures in the outer sheath would be visible provided all-round coverage were made. Kinks and dents would also be readily identified.

Inspection from some distance by ROV or diver would also enable abnormalities in the catenary shape of risers to be observed. By comparing the configuration which the riser takes up, with previous observations and the results of laboratory tests, any deviation in geometry outside predetermined limits would signal the requirement for more detailed inspection.

Problems would be encountered on risers at bend restrictors and in areas where buoyancy is attached, the latter being an area of particular interest since the buoyancy attachment system may itself be the source of external sheath damage. The method would not be applicable for buried flowlines, the underside of unprotected lines or for risers encased in J-Tubes. (Figure 1). The presence of marine growth would also inhibit inspection.

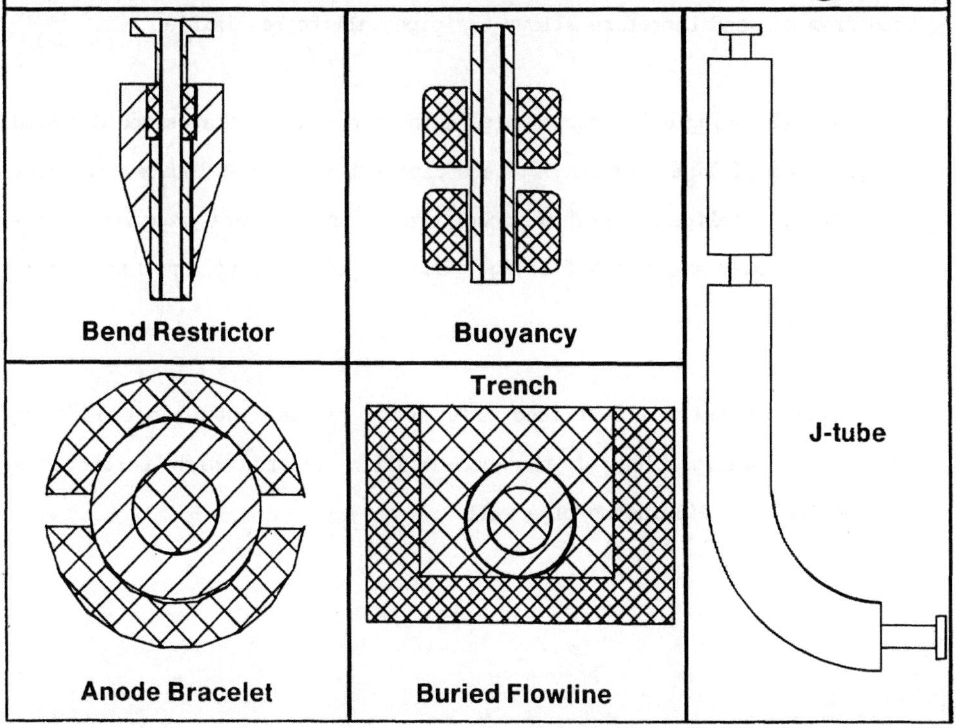

OBSTRUCTIONS TO INSPECTION Figure 1

External calipering could also be performed by diver, ROV or by a purpose built crawler. The same access limitations at bend restrictors, areas of marine growth, buoyancy and on protected flowlines would apply.

It is suggested that such calipering equipment should be capable of identifying variations in diameter greater than 5% which extend for 20% of the pipe circumference.

Internal visual inspection requires a compact camera and carrier system, a suitable means of driving the device, and will require an interruption to normal production. In certain cases the pipe would have to be flushed and cleaned to ensure adequate image quality. Existing downhole technology would most probably be applicable. Video cameras have been used to inspect well completions. The use of a clear gel in front of the camera which is withdrawn on wireline system creates an area in front of the lens which is free from oil and therefore allows the pipe wall to be visible.

Internal calipering using cupped pigs with thin gauging plates would reveal the presence of liner damage or distortion but would not locate it. More sophisticated calipers could be incorporated on a camera carrying device which would also require a "footage counter" to pinpoint the location of any feature.

The inspection methods described under this category would not identify damage or deterioration of the non-metallic layers which perform the sealing function within the pipe wall.

Second Pass Methods

Second Pass Methods for inspection include techniques which are capable of determining, in-situ, the condition of one or more of the constituent layers in the pipe wall. This will include detection of wear, cracks or distortion. These methods will be relatively fast in operation and should ideally work on the principle of "Inspection by Exception". That is to say that a characteristic "signature" for a good, or new, pipe is compared with the signal measured on the subject pipe and areas where differences or anomalies appear are highlighted for closer inspection.

The techniques which are considered applicable to this group include:- Eddy currents, ultrasonics, thermography, mechanical excitation, acoustic emission and TV Holography.

All of these techniques are currently used as NDT methods in a variety of industries. However their effectiveness in flexible pipe applications is governed by the composite construction of the pipes and the "masking effect" that each layer has on the others. (Figure 2).

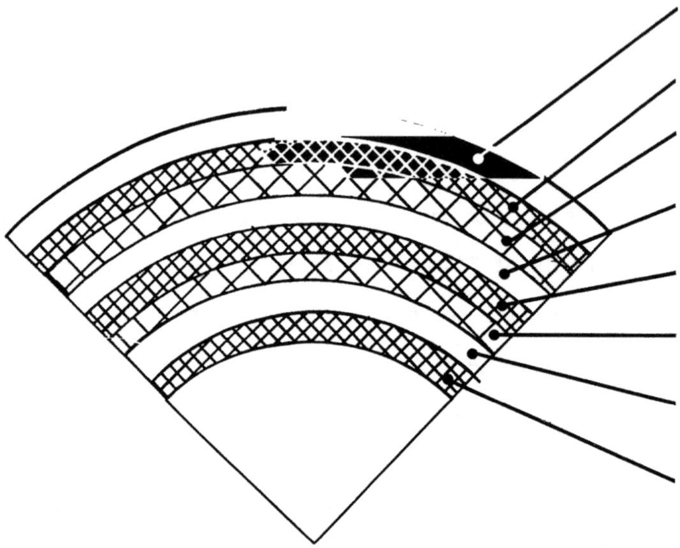

Construction comprises alternate steel and polymeric layers
(Unbonded pipe example)

Figure 2

Second Pass methods potentially offer the most valuable information from in-situ monitoring and should therefore be given a high priority in the development of NDT methods.

The metallic components of pipe construction naturally lend themselves to magnetic and eddy current methods which are perhaps the most promising techniques currently being considered. They have the advantage that direct contact between sensor equipment and pipe is not always necessary and that they can be used in a "dynamic scanning" mode. That is to say, in operation the sensor is moved in relation to the pipe, thus offering relatively high inspection speeds.

When inspecting homogenous steel components, such as a steel pipe wall the presence of cracks or corrosion will be evident since they cause a disturbance to an otherwise uniform field. Flexible pipes, being constructed of numerous elements, have large numbers of discontinuities and section changes and thus the sensitivity of these methods is likely to be impaired. Also the products of corrosion or abrasion will still remain encased within the structure and may impair sensitivity. Additionally, the presence of thermoplastic and elastomeric layers imposes a minimum stand-off distance between sensor coil and steel members. The sensor coil is therefore operating in a region of reduced magnetic flux and sensitivity is reduced accordingly. These methods are used for wire rope inspection devices and with a combination of "Hall" effect transducers and passive sensor coils it is possible to detect both material wastage due to corrosion and breaks in single strands in mooring lines.

For each pipe size and design it is envisaged that the magnetic field strength and detector coils will have to be optimised.

Experiments have been performed by BP at Sunbury using an existing wire rope tester, which has been used on the mooring lines of BP's Buchan Alpha semi-submersible production vessel. This device is capable of detecting a break in a single wire 2mm diameter in a cable of overall diameter 72mm.

The test pieces comprised a 2" low pressure rubber hose with brass coated steel braid reinforcement (strand diameter 0.4mm) and a 1" 5000 psi unbonded pipe with armour wires 2mm x 4mm in cross section.

Additional wires of varying diameters were attached to the inside and outside of the test pieces and the wire rope tester passed over the test area. The aim was to ascertain whether breaks in the additional wires could be detected in the presence of the reinforcement members of the pipes. (Figures 3a and 3b).

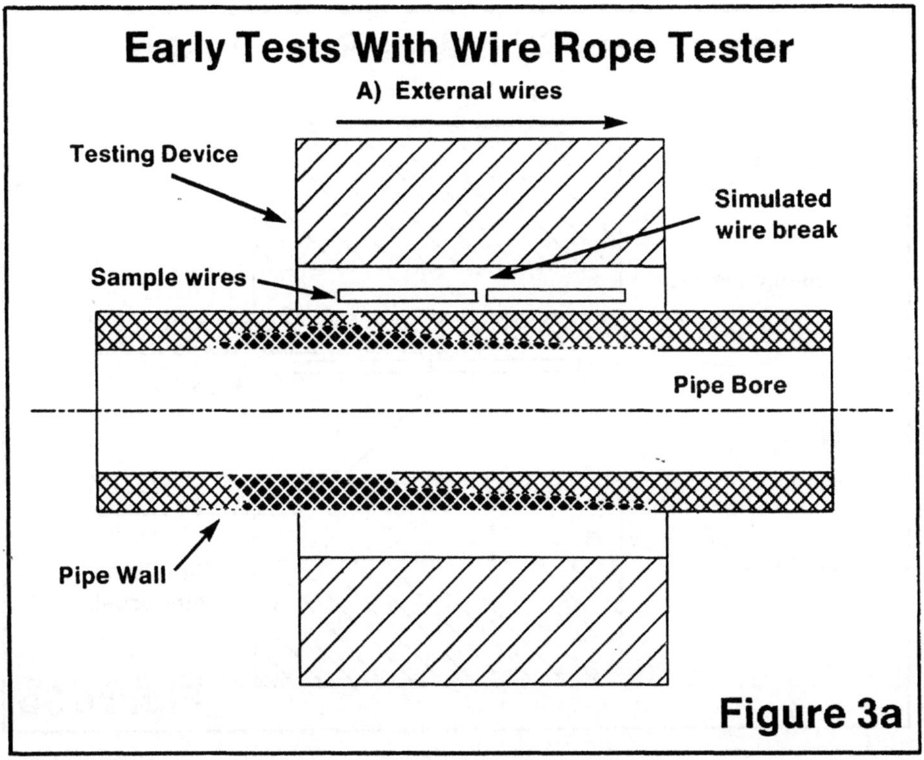
Early Tests With Wire Rope Tester
A) External wires
Testing Device
Simulated
wire break
Sample wires
Pipe Bore
Pipe Wall

Figure 3a

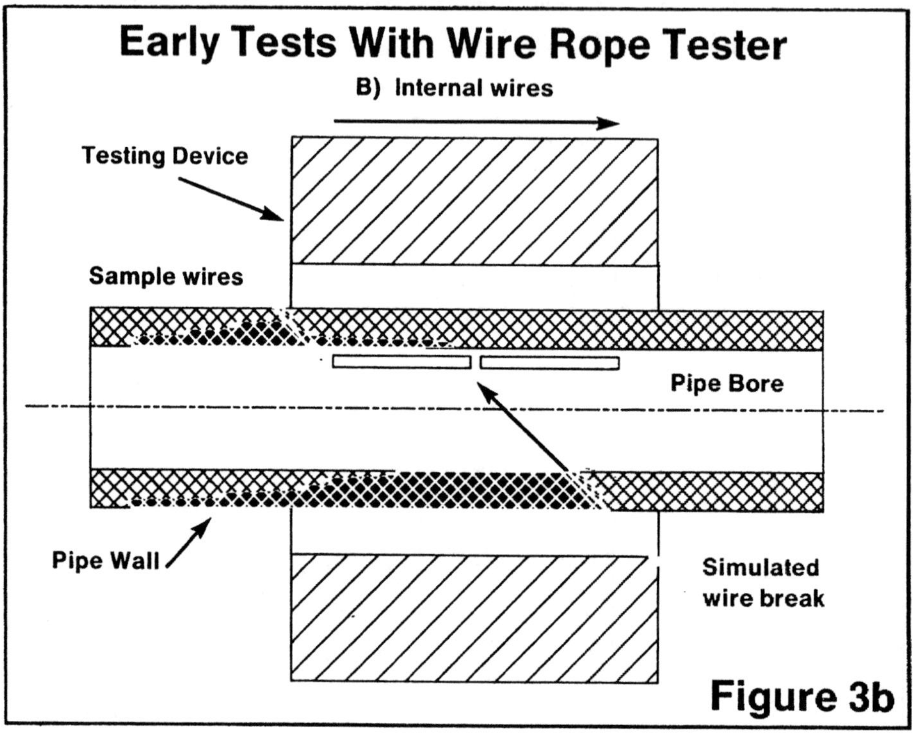

Figure 3b

For the bonded hose, breaks in 0.7mm diameter wires on the outside and 1.4mm diameter on the inside of the sample were detected.

For the unbonded pipe breaks in 2.5mm diameter wires were detected. Results from internal wires were inconclusive but this is expected to be due to the experimental conditions. (Constant speed of sensor head must be maintained).

Ultrasonic techniques at first appear to offer some promise since they are used for thickness measurements and can detect loss of wall thickness in steel components. Time of flight diffraction methods are also being used to detect internal cracks in steel. (Reference 1).

For flexible pipes however the many interfaces between adjacent layers create reflections and loss of sound energy. For example, approximately 90% of sound energy is reflected on meeting a steel surface in air. Hence the possibility for looking 'through' steel layers is very limited, particularly the interlocked carcass which is effectively four thin layers of stainless steel. (Figure 4).

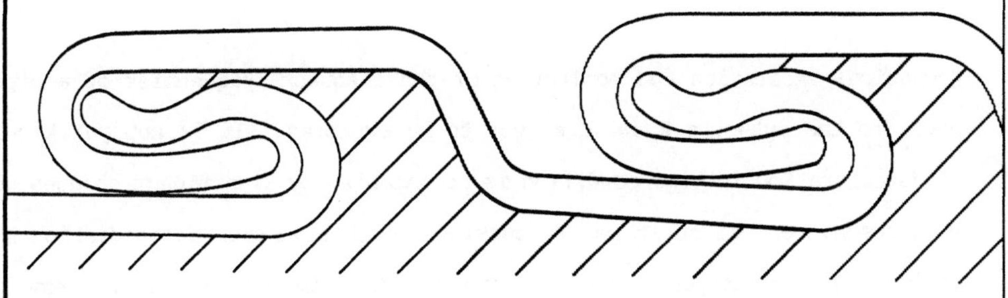

Section Through Interlocked Carcass

Figure 4

There may however be scope for using ultrasonics to inspect the inner liner surface for abrasion from inside the pipe by relying on reflected energy or for detecting delaminations in bonded pipes. This may obviate the need to flush and clean the pipe bore as suggested for visual internal inspection.

Thermographic methods are used successfully for inspection of fibre reinforced composites and will detect delaminations.

The potential for application to flexibles is reduced due to the steel components conducting heat axially along the pipe and due to the extremely poor propagation of infra-red in sea water. This has been borne out by BP's own laboratory experiments. There may however be limited scope for use near riser top connections and where the pipe is used in air.

Mechanical excitation and monitoring of the resonant frequencies of a pipe have, to the author's knowledge, yet to be explored. It is not yet known if deterioration in pipe condition is accompanied by significant changes in pipe resonance. If so it may be possible to observe such deterioration by monitoring the response to a known input stimulus, or even from a quasi-random stimulus such as wave action, for a riser system.

Acoustic emission techniques are being addressed by one proposed Joint Industry Programme for unbonded pipes. It is apparent that acoustic noise generated by fretting of adjacent steel components can be detected and the onset of seizing can be observed. It is believed to be possible to mount sensors on the outside of the outer sheath and to achieve up to a 1 metre coverage of the pipe length.

TV holography is a technique which compares the holographic images of the same object in two different loading conditions. Computer image enhancement is used to provide clear results. A typical application is vehicle tyres which are viewed in the pressurised and unpressurised condition (Reference 2). Areas of excessive distortion are easily identifiable. No experiments have yet been done on flexible pipe.

Third Pass Methods

Once a potential flaw or damage in a pipe has been identified by either first or second methods the decision would be made whether to carry out further examination in situ or to take the unit out of service and perform a detailed examination in controlled shore based conditions. The damage may, of course, be so great as to render the pipe unusable but, subject to detailed inspection, the option of removing the damaged section and introducing an intermediate flanged connection may be considered.

The methods which would be employed comprise those techniques which, while potentially capable of detecting small flaws in individual layers are either very slow in operation or rely upon painstaking post-inspection data manipulation by the operator. In addition it is likely that the equipment required is not suitable for offshore or subsea use.

These techniques include the full range of radiation methods such as X-Ray and Gamma-Ray radiography and tomography. Tomography is a method of constructing cross sectional images from a large number of single radiographic images taken across the sample as the sample rotates through 360° relative to the source. The same technique is used in whole body scanners for medical purposes.

It is estimated that a pipe could be scanned at a rate of 1 metre per hour to a resolution of 1mm, thus demonstrating the limitations for in-situ application. BP and Department of Energy are currently conducting a programme of X-Ray tomography inspection of pipe samples with artificial flaws.

These third pass techniques are likely to provide the most detailed information of all the methods discussed but the speed of operation and coverage is extremely slow and it is only practicable to adopt them for inspection of areas where flaws are known to exist, or where loading conditions are known to be high, ie at riser top end connections, areas in splash zone and adjacent to mid water arches etc.

It should however be mentioned that these techniques could prove valuable as quality control methods during manufacture and pipe suppliers should be encouraged to give this consideration.

Thus we have an outline strategy for condition monitoring which comprises:-

- Periodic visual and dimensional checks - FIRST PASS INSPECTION followed, less frequently by

- Induced flux techniques which will locate and partially quantify non-visible flaws, mainly in steel layers - SECOND PASS INSPECTION, followed finally by

- Pipe removal and detailed local examination to determine extent of damage using radiographic or other means - THIRD PASS INSPECTION.

NON METALLIC MATERIALS

In addition to the above philosophy, due consideration must also be given to the range of materials used in pipe construction.

Most of the physical principles already discussed are suitable for inspecting metal components but the main pressure containing layers in the pipes are either elastormers or thermoplastics.

These materials do not readily lend themselves to NDT. Indeed, nearly all tests to determine material properties are destructive in nature and use small coupons of material in laboratory scale equipment. Once again, the encasement of the materials within the pipe wall further complicates any inspection undertaken.

Proposals have been put forward for use of electrical properties to detect reductions in layer thickness (Reference 2). These use the steel armours and sea water as conductors either side of the outer sheath. Periodically a low frequency A.C. current is applied and a detection circuit used to measure the impedance of the system. An assessment of sheath wastage is made based upon any change in impedance compared to an initial datum figure.

The authors are not currently aware of any practical tests which have been performed to evaluate this technique.

Turning to the inner, pressure layers, the operating environment for hydrocarbon duties will invariably comprise pressures up to 5000 psi or higher, temperatures up to 90 or 100°C, exposure to the transported fluids and, for dynamic riser applications, cyclic bending.

Whilst these levels of pressure and temperature are of little consequence to the metallic components they can and will, have an ageing effect on the elastomers and thermoplastics. This can include loss of tensile strength, increase in brittleness and porosity, and for bonded pipes a possible reduction in bond strength. Finally epoxy resins, which are used inside the end fittings, are subject to softening and loss of strength when exposed to elevated temperatures.

All of these properties are included in the functional requirement for the materials in the pipe application, but there are no means of assessing these properties in-situ.

One solution is to place a number of sample coupons in the flow stream and to remove some periodically for destructive testing in laboratory equipment.

This requires a large number of coupons and does not give a full representation of the actual service conditions. For example the pipe wall will be subject to a pressure differential which is constantly driving gas through the material. The pipe wall will be in a stressed condition due to pipe internal pressure, cyclic bending stresses in the case of risers, and frictional effects from adjacent layers.

Alternatively short spool pieces of pipe could be installed in each line at a strategic location where pressures, temperatures, dynamic loads and internal abrasion are determined to be most severe, and these could be removed and inspected destructively after a pre-determined interval. The major drawback with this scheme is the high cost of spool pieces since they will of necessity incorporate expensive end fittings, and the inability to obtain cumulative life deterioration, (unless a multiplicity of spools is used).

IMPLANTATION METHODS

Returning to the principle of outer sheath impedance measurement. This represents a different approach from intervention by a specialist tool. It requires some local modification of the pipe design to enable the electrical connection to the armour wires to be made. (It is a matter of detail whether this is a permanent connection or one which is made only for periodic measurements).

The concept of altering pipe design and construction to facilitate inspection, or perhaps more accurately, condition monitoring, is worthy of some discussion. Another patent (Reference 3) suggests the incorporation within the pipe wall, of an electrical cable which forms part of a pilot circuit. In the reference application it is envisaged that a break in the pilot circuit caused by excessive extension of the cable would trigger the closure of a suitable shut down valve.

This system would rely upon the pilot circuit cable being sufficiently robust as to withstand the obvious fretting to which it would be exposed within the pipe wall, but at the same time, designed to break before the pipe itself ruptures. It would serve mainly as an early warning of excessive tensile loads. Once activated, it could not be re-used.

Furthermore, the problems of introducing an additional element into the pipe during manufacture are not insignificant. The simplest means is likely to be the attachment of a localised pilot cable in the vicinity of a riser top end connection prior to the installation of a bend restrictor. This would at least offer monitoring of the region of the pipe which will experience highest tensile loads.

Nevertheless, the reliability of implantation methods would have to be carefully considered and they may serve, at best, as a "flag" to initiate the mobilisation of other NDT methods.

CONCLUSIONS

There is a real and growing need for workable NDT methods for flexible pipes.

The nature of pipe construction and the variety of materials used means that no single inspection technique will reveal the condition of all pipe components.

An inspection strategy is required which commences with visual external and/or internal inspection leading to less frequent but more quantitative measurements to highlight areas for more detailed examination, possibly following removal from service.

Methods for assessing the condition of non-metallic materials, in situ must be found to avoid the need for numerous in-stream coupons or sacrificial spools.

In built sensor circuits may be feasible but their reliability is likely to be low and they will have an adverse effect on pipe cost.

In concluding this presentation it is fair to say that any paper which is entitled "The problems with ...." is certain to include some negative observations. However it is our opinion that the subject of NDT of flexible pipes represents the sort of challenge which will eventually bring out some innovative solutions and we look forward to seeing some of these reported at future conferences of this type.

ACKNOWLEDGEMENTS

The authors wish to thank the British Petroleum Company plc for their kind permission to present this paper.

Thanks are also due to the Deparment of Energy Petroleum Engineering Division, for their co-operation in feasibility studies and testing.

REFERENCES

1. Subsea Ultrasonic Inspection Using Advances Imaging Techniques. W. Browne. SCT Conference "Advances in Underwater Inspection and Maintenance. May 1989.

2. Critical review of recent developments in electronic speckle pattern interferometry, J. R. Tyrer, Holographic NDT. Proc SPIE Bellingham, Washington 604 (1986) pp 95 - 111.

3. UK Patent Application G.B.2180073A August 1985

4. Patent Application G.B.2005036A Sept 1978

Part III

Pipeline Design, Fabrication and Installation

PIGGABLE PIPELINE WYE CONNECTION - DEVELOPMENT AND DESIGN

Presented by K. McKay

Authors - P.G. Brown, J. Ritchie, K. McKay,
J P Kenny Caledonia Ltd.
A.J. Grass,
University College London

ABSTRACT

This paper considers the design and development of a 30-inch piggable subsea pipeline Y-connection which a major operator intends to install in a central North Sea gas pipeline. A Y-connection of this type enables two pipelines of the same diameter to be tied-in whilst maintaining full piggability. This will be the first known fully piggable Y-connection for a subsea gas pipeline and as such offers significant advantages over other tie-in methods, such as tee connections and riser platforms, both of which require some degree of intervention.

General aspects relating to the overall arrangement such as valving, piping layout requirements and protection and installation requirements are discussed although emphasis is

Volume 24: C. P. Ellinas (ed.), Advances in Subsea Pipeline Engineering and Technology, 207–227.

placed on the Y-connection itself. The major stages in the design and development of the Y-connection are explained. These include design of the internal geometry to allow unhindered passage of pigs, structural analysis and material selection. A method developed for predicting minimum flow velocities required in the pipeline to achieve successful pigging is also presented.

INTRODUCTION

The concept of a fully piggable pipeline Y-connection can be considered to be relatively recent, with the first such example installed in an oil/condensate line in the Norwegian sector of the North Sea in 1985. The need for a tie-in method which permits pigging through both tied-in pipeline branches with no intervention or weather dependency has arisen due to the cost and complexity involved in utilising other solutions. These solutions are generally constituted by either riser platforms or subsea 'tee' connections. Riser platforms by their nature are very costly and require intervention whenever any pigging takes place. Considering piggable subsea 'tee' connections, two main types are currently in operation: a 'tee' with removable subsea pig receiver which requires intervention and is weather dependent and a 'tee' with a return pigging loop tied back to the parent platform. Figure 1 presents illustrations of the three tie-in methods described.

The economic viability of a Y-connection depends to a large extent on its location within a system. The capital cost of a wye is justifiable where an optimum pipeline tie-in position is located in an area remote from suitable existing platform facilities. Since all branches of the Y-connection must be the same diameter for pigging purposes, viability is also dependent

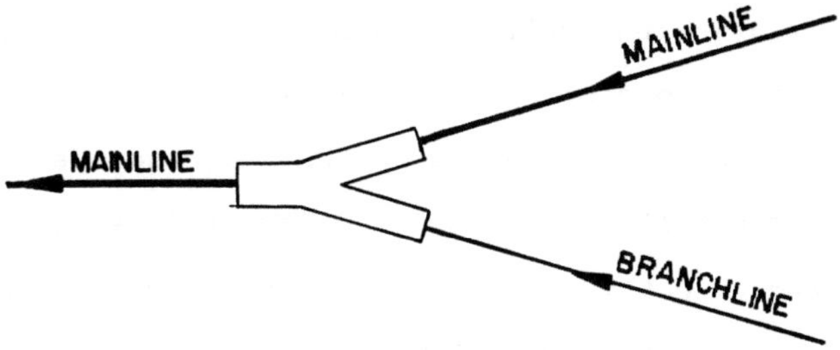

1) Y-CONNECTION

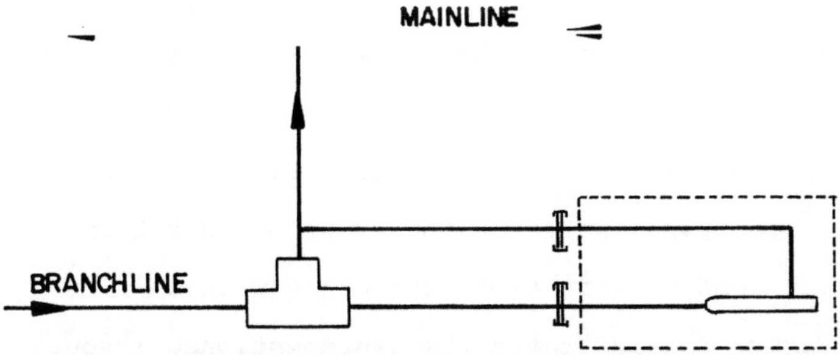

2) TEE CONNECTION WITH REMOVABLE SUBSEA PIG RECEIVER

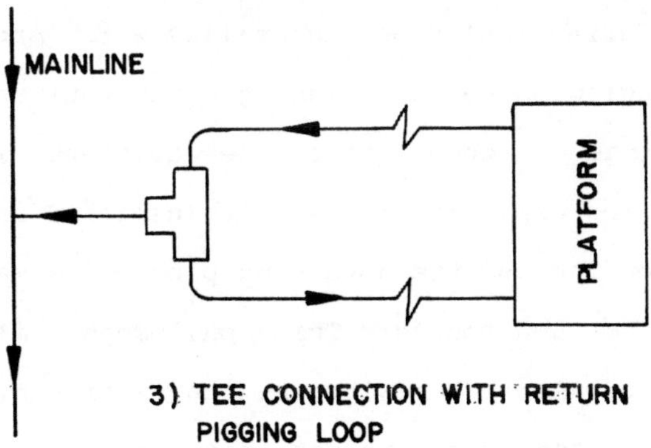

3) TEE CONNECTION WITH RETURN PIGGING LOOP

FIGURE 1
TIE-IN METHODS

on sufficient flow being available through the secondary branch of the Y-connection to ensure a suitable flow regime.

In the correct location the potential benefits of a Y-connection are substantial. These benefits include the ease of pigging, capital cost savings over other tie-in methods, low maintenance costs and a simpler and more compact overall layout.

The Y-connection itself is either a cast or forged steel component and is housed subsea in a supporting tubular steel framework assembly. The Y-connection piping assembly generally comprises of two in-line ball valves on each upstream branch, although this may be varied depending on specific commissioning and isolation requirements. The piping system also includes bleed valves, pipeline expansion loops on each branch of the Y-connection and a protection structure with an integral support skid. Figure 2 illustrates the aforementioned layout. Where the Y-connection is located in an area of fishing activity, the associated protection structures are generally required to be steel framed, 'deflection' type assemblies which provide a low profile non-snagging shape for fishing equipment, anchor wires and so on. It is intended that the complete Y-connection assembly, including expansion loops, is installed in a single lift, which depending on the governing pipe size may range in weight from 100 to 600 tonnes. This philosophy will effect a saving on the costs associated with the subsea tie-in.

It should be noted that although this paper describes work undertaken on a 30-inch diameter Y-connection the principles involved may be readily adopted to other diameters.

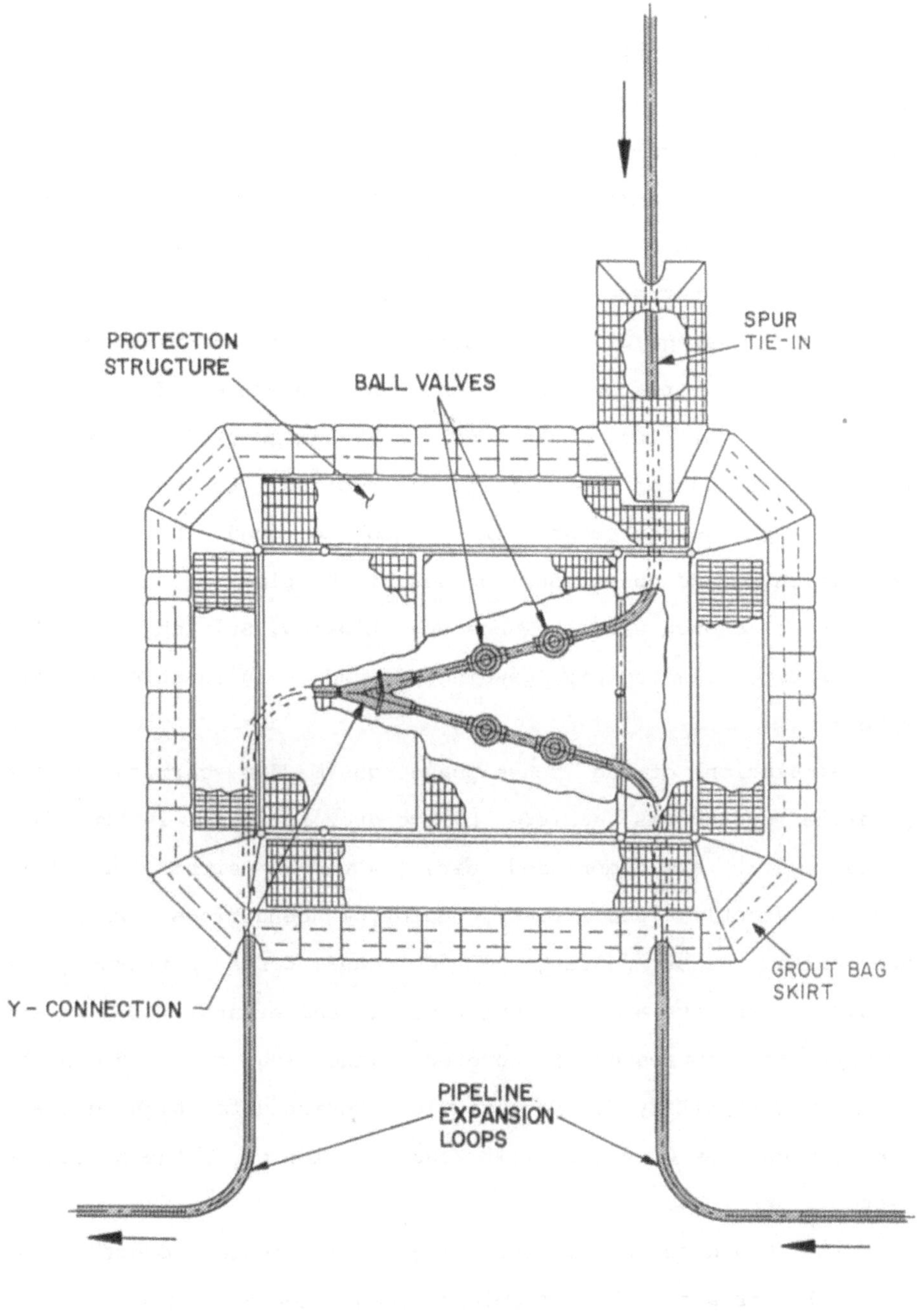

FIGURE 2

TYPICAL Y-CONNECTION ASSEMBLY LAYOUT

INTERNAL GEOMETRY DESIGN

Testing Programme

In order to determine an optimum internal geometry for the Y-connection a programme of scale model testing was undertaken at a commercial research centre. The aim of this programme was to develop an internal profile and geometry which would allow unhindered passage of a representative range of pigs over a variety of flowrates and conditions. At the outset of the test programme no detailed pigging philosophy had been developed for the pipeline system therefore it was necessary to identify those generic pig types most likely to be used in both the commissioning and operational phases of the pipeline's life. The pig types chosen were: Tuboscope Linalog, British Gas OLI, caliper pig, cleaning pig, gauging pig and an inflatable pipeline sphere.

Simulation of the dense phase gas medium carried in the pipeline system was achieved by conducting tests in two flow media, water and compressed air. Although neither of these fluids exactly reproduces prototype conditions with the pressurised dense phase gas, the actual tests performed are considered to give a good simulation of real events. The tests in water and compressed air covered a wide range of fluid density and compressibility characteristics designed to highlight any special features of pig mobility characteristics in the prototype system.

It should be noted that pigging of liquid product lines generally presents less problems than pigging of gas lines. There is substantial freedom from velocity excursions and the wear on the pig is usually reduced in liquid lines. Therefore, the findings of this investigation will be equally applicable to Y-connections in liquid product lines.

Two test rigs were constructed to carry out the experimental investigation programme. These were based on 4-inch and 8-inch pipeline systems, providing a diameter scaling to the 30-inch prototype pipeline of 1 to 7.5 and 1 to 3.75 respectively. In the 4-inch rig, for practical purposes water was employed as the driving fluid with flow velocities ranging from 0.01 to 1.5 m/s, whereas the 8-inch rig utilised compressed air with velocities ranging from 1 to 4.3 m/s.

For the 4-inch tests, eight model Y-connections were constructed from perspex, each with a different internal geometry. The variables employed in the internal geometry were as follows:-

- angle between inlet branches;
- radius between inlet and outlet branches;
- inlet branch intersection radius; and
- increase in internal diameter.

Figure 3 gives a definition of these terms.

Initially seven 4-inch diameter Y-connections were tested, the eighth wye connection was constructed after gaining feedback on the performance of the others.

A large number of test runs were conducted over the range of pig types, internal geometries and flowrate conditions, resulting in a predicted optimum internal geometry. This geometry was then reproduced at the larger scale and tested in the 8-inch compressed air test rig. A number of test runs were carried out over a range of parameters and the results confirmed the

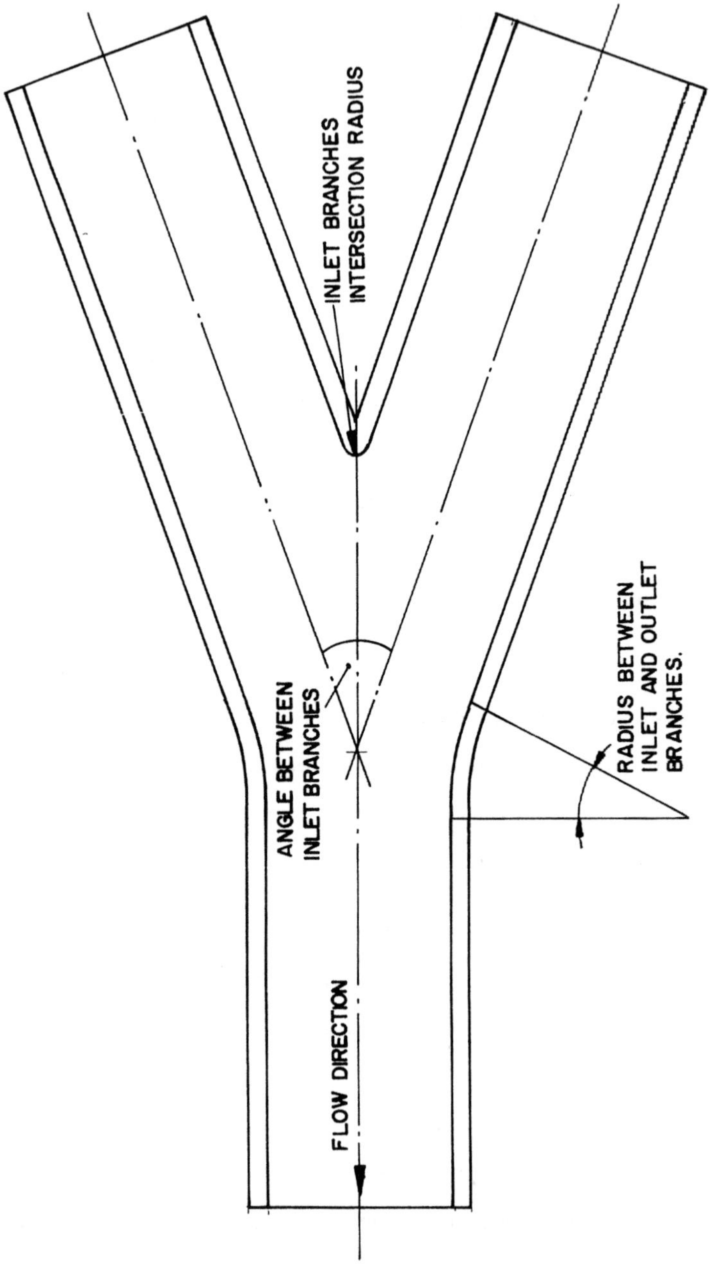

FIGURE 3
DEFINITION OF Y-CONNECTION TERMS

conclusions drawn from the 4-inch tests.

The tests showed that the most critical condition is when the pig is stopped in the wye and flow is restarted in one upstream branch only. The flow velocity required to restart the pig, the 'threshold velocity', is then set as the minimum flow velocity in the pipeline system.

Scaling Criteria

Model pigs

The modelling criteria necessary to satisfy complete dynamic similarity with prototype behaviour, are extremely complex. This situation is exacerbated by the compressibility combined with the high density of the pig transporting fluid in the prototype. The kinetic energy of a moving pig emerging from the sealed condition in the approach pipeline helps to carry the pig through the enlarged Y-connection. Logical evaluation supported by mathematical modelling of the pig dynamics in the Y-connection indicates that the critical minimum velocity at which the pig is arrested in the Y-connection is less than the restart threshold velocity discussed previously. Prediction of the latter restart velocities using the model test results was therefore the essential priority of the investigation. Careful observations of the dynamic behaviour of the various pig geometries were carried out however to ensure smooth progress of the pigs through the Y-connection at the higher operational velocities.

A set of model pigs was therefore manufactured with mass simply scaled down assuring constant density and hence using the criteria:

$$\frac{M}{d} = \text{constant} \tag{1}$$

Hence: $$\frac{M_m}{M_p} = \left(\frac{d_m}{d_p}\right)^3$$

Where (M_m, d_m); (M_p, d_p) are respectively the pig mass and pipe diameter for the model and prototype.

In order to more fully explore the range of possible dynamic effects relating to the pigs' movement through the Y-connection, a further set of model pigs were manufactured using the following mass scaling criteria:-

$$\frac{M}{\rho d^3} = \text{constant} \tag{2}$$

where ρ is the transporting fluid density.

Hence $$\frac{M_m}{M_p} = \frac{\rho_m}{\rho_p} \frac{d_m}{d_p}^3$$

The ratio in equation (2) essentially represents the momentum of the pig relative to that of the surrounding fluid. This parameter is thus clearly important in influencing the acceleration and deceleration behaviour of the pig during periods when its velocity changes relative to that of the surrounding fluid. In the case of the compressed air tests in the 8-inch facility, equation (2) produced extremely light weight pigs capable of more realistic dynamic response in the air flow. These two different scaling criteria therefore yielded a wide parametric range of test conditions. This was considered important to reveal any peculiar dynamic behaviour in the pig motion which might be of practical significance in the full scale system.

Prototype threshold velocity prediction

The model tests showed that hold up of the pigs in the Y-connection would be avoided if the friction between the pig and the walls of the Y-connection is minimised. If this objective is achieved then the remaining source of resistance to the motion of the pigs would be limited to the gravity induced friction with the bottom of the Y-connection.

This frictional resistance F_f, is proportional to the pig weight and is given by:

$$F_f = f\, M_s\, g \qquad (3)$$

where: f = Friction coefficient

M_s = Pig 'submerged' mass = $\left(1 - \frac{\rho}{\rho_{pig}}\right)$

M = Mass of pig

ρ = Density of fluid

ρ_{pig} = Density of pig material = $\frac{\text{Mass}}{\text{Volume}}$

g = Acceleration of gravity

If the pig stops in the wye due to fluid by-pass and loss of cup seal drive, there is still a reduced fluid force acting on the pig. This force can be equated to the net rate of momentum loss of fluid, which is an approximate proportion of the rate of fluid momentum flux in the pipe.

The fluid driving force is therefore $F_D = \beta . \rho . A . V^2$ (4)

where: β = Constant of proportionality

ρ = Density of fluid

A = Cross sectional area of pipe

$= \pi \frac{D^2}{4}$ where D = pipe I.D

and V = Mean velocity of fluid

The constant of proportionality, β, is not only dependent upon the type of pig, but also on the location in the Y-connection where the pig/sphere is stopped. The effect of pig location on the minimum fluid velocity required to free the pig has been the subject of a detailed investigation.

By equating the driving force to the frictional resistance the minimum fluid velocity necessary to initiate the pig motion (threshold condition), can be calculated.

$$\text{Now} \quad \frac{F_F}{F_D} = \frac{fM_Sg}{\beta \rho A V^2} \tag{5}$$

For simplicity, β will be assumed constant for a given pig. Therefore for dynamic similarity,

$$\frac{f_m M_{S,m}}{\rho D^2{}_m V^2{}_m} = \frac{f_p M_{S,p}}{\rho_p D^2{}_p V^2{}_p}$$

$$\text{or} \quad \frac{V_m}{V_p} = \left[\left(\frac{f_m}{f_p}\right) \left(\frac{M_{S,m}}{M_{S,p}}\right) \left(\frac{\rho_p}{\rho_m}\right) \left(\frac{D_p}{D_m}\right)^2 \right]^{\frac{1}{2}} \tag{6}$$

Subscripts m and p refer to model and prototype, respectively.

The above relationship was then used with reasonable certainty to predict the prototype threshold velocities for each pig type, except the sphere.

In the case of the sphere, the predicted prototype threshold velocities showed a marked discrepancy between the light and heavy model results based on the scaling criteria in equation (6). This indicates that the mechanism by which the sphere is transported through the Y-connection is not correctly modelled by equation (6). The discrepancy arises from the fact that the sphere is subject to complex rolling rather than simple sliding resistance.

A further series of tests involving spheres showed that the force required to move the sphere is not directly proportional to its mass. It was further shown that an exponent of 0.6-0.7 should be applied to the mass term in equation (6). This then enabled the prototype threshold velocity for the sphere to be predicted.

The testing programme served to optimise the internal geometry of the Y-connection away from the original concept of combining three constant bore pipe sections. This optimisation of the internal geometry enabled the structural analysis to begin.

STRUCTURAL ANALYSIS

As noted above, the starting point for the structural analysis is the internal geometry derived from the testing programme. External loading on the Y-connection was minimised as far as possible by means of the pipeline expansion loops and clamping arrangement within the assembly piping. Figure 2 shows details of the Y-connection assembly. However, due to the large pipeline expansion forces present, significant external loadings are still transmitted through to the Y-connection. In addition, internal pressure loadings arising from both the operational and hydrotest conditions were also substantial.

Two Dimensional Analysis

A two dimensional finite element analysis of various sections of the Y-connection was conducted in order to establish approximate wall thickness and stiffener requirements. This methodology was adopted as an initial step prior to undertaking a more complex three dimensional analysis.

Initially, the full spectrum of loadings were reviewed in order to establish the most critical loading combination. The hydrotest pressure case was concluded to be the most severe loading case.

For the two dimensional finite element analysis, the 'ANSYS' finite element (FE) program was used, with particular emphasis being placed on the usage of isoparametic stress solid elements. All completed analyses were performed using linear elastic stress criteria. The ANSYS preprocessing mesh generation capability was used extensively to generate model nodes and elements. Mesh refinement was checked by verifying simple model analyses against empirical formulae, with the mesh refinement exercise also being utilised to check the sensitivity of results.

Two methods of establishing the wall thickness requirements were considered during the analysis:-

- Use of an iterative approach to obtain a uniform wall thickness which would accommodate the hydrotest loading. This involved analysing consecutive axial and longitudinal material sections in the vicinity of the Wye intersection, and refining dimensions at each stage.
- Use of a central fin stiffener and ring stiffener, to provide sufficient material to minimise material deflection under the hydrotest loading. The axial and longitudinal material sections were analysed as before, but with most of the pressure load reacted at the stiffeners.

The former approach resulted in an estimated Y-connection weight of 25-30 tonnes, which was considered to be unacceptably high, hence the stiffener approach was pursued. Subsequent discussions held with potential manufacturers indicated that a maximum weight of around 20 tonnes should be aimed for.

Maintaining the Y-connection weight within this limit widens the range of potential manufacturers and minimises physical constraints with respect to the manufacturing process. The central stiffener geometry was further refined until a suitable maximum deflection was achieved under hydrotest load. This required a stiffener height of around 700 mm, which was considered reasonable. Suitable contouring of sharp edges was then incorporated into the model to reduce stress concentrations, and further analyses conducted to determine stress levels. The maximum equivalent stress level obtained was approximately 1.45 times maximum allowable stress. However, as a two dimensional analysis invariably overestimates stress levels, it was decided to utilise this geometry to initiate the three dimensional analysis.

Three Dimensional Analysis

A series of three dimensional finite element analyses were conducted in order to determine the maximum stress levels within the material of the Y-connection.

The analyses again utilised the 'ANSYS' FE program, this time with the emphasis being placed on the usage of three dimensional isoparametric stress solid elements were employed. All analyses were linear elastic, and maximum usage was made of a preprocessor mesh generation program to generate nodes and elements. It was considered unnecessary to verify the mesh sizes, as the previous two dimensional analysis verification was still applicable. The loadcase adopted was the hydrotest pressure with related external loads .

The finite element model was generated as a quarter section and reflected twice, as double symmetry exists, to form a 360° body. The model is presented in Figure 4. The external loads were applied as displacements at the branch ends of the Y-connection.

During the analysis a number of stiffener configurations were adopted and the results showed that the configuration shown in Figure 4 was the most acceptable: giving a maximum equivalent stress which marginally exceeded the maximum allowable stress of 403 MPa. This stress result was subsequently refined by a sub-model analysis which brought the stresses within the acceptable limits.

Deflections within the body of the Y-connection are a maximum at the central stiffeners. Here, a deflection of 2 mm was calculated, which is considered acceptable.

It was considered necessary at this stage to look at the branch intersection area in more detail, as the element mesh on the three dimensional model could not give the necessary accuracy. A sub-region analysis was therefore undertaken around the area of highest stress.

Detailed Sub-Model Analysis

A detailed sub-model analysis was conducted to determine the stress levels within the Y-connection branch area more accurately.

The area modelled during this stage of the analysis was a quarter section and was restrained using a 'sub-region' feature

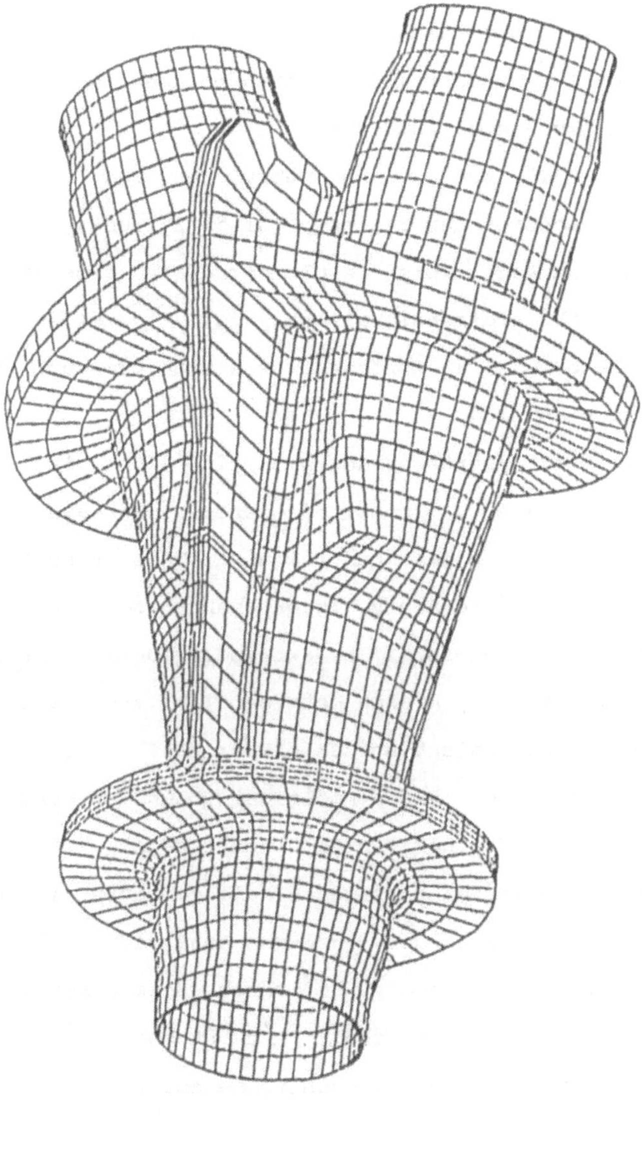

FIGURE 4
THREE DIMENSIONAL FINITE ELEMENT MODEL

of the ANSYS program. This allows reactions at the same locations in the main three dimensional analysis to be applied to the sub-region model. These reactions then became the boundary conditions for the sub-model analysis.

Analysis of this sub-model geometry resulted in stress levels at the Y-connection centreline of slightly less than the maximum allowable stress. This result correlated closely with the results of the three dimensional analysis. Edge effects were noted within the sub-model at surfaces where both pressure forces and restraints were applied.

These edge effect stresses do not exist in reality, but within the model cause a localised stress increase, which diminishes towards the model centreline. This results in reported stresses being moderately higher than those present in the actual material. Therefore, by extrapolation it was shown that stress levels in the branch intersection area will not exceed the allowable equivalent stress of 403 MPa.

As a final check on the symmetrical boundary conditions imposed on this model and other finite element models, the sub-region geometry was reflected to produce half symmetry in place of quarter symmetry. This larger model was analysed and stresses found to correlate very closely to those determined in the quarter symmetry model on the relevant boundary. This boundary condition was therefore proven to be accurate.

Conclusion

As a result of the analyses undertaken, the following conclusions were drawn:-

- Stresses within the Y-connection, with optimised geometry, will not exceed the maximum allowable equivalent stress of 403 MPa.
- Deflections within the body of the Y-connection were greatest at the centreline of the longitudinal stiffener, and were in the order of 2 mm. These were not considered significant with regard to structural integrity or operational performance.
- The hydrotest loadcase utilised during this analysis constituted the most severe loading condition the Y-connection would encounter; stresses induced during normal pipeline operations would be less than those mentioned above.
- The optimised Y-connection geometry may undergo minor modification to accommodate fabrication process requirements once details of fabrication methods are finalised.

MATERIAL SELECTION

The major material requirements for the steel used in Y-connection manufacture are as follows:-

- Material must be readily weldable to linepipe material (API 5L X65 modified).
- Due to the severe, sour nature of the dense phase gas product NACE sour service requirements must be met, sulphide stress cracking and hydrogen induced cracking tests must be performed.

- Minimum allowable yield strengths will be 403 MPa in branch intersection area, 320 MPa in other areas.

Manufacturing Methods

The most appropriate methods for Y-connection manufacture are considered to be casting and forging. Each method has its own relative merits and any decision on selecting one or other method is dependent on numerous general and project specific factors. Due to the innovative techniques involved in Y-connection design and manufacture it is considered necessary to manufacture a 'pilot' Y-connection which would be hydrotested and then used for destructive testing purposes to ensure that the production Y-connection achieves the required strength and corrosion resistance levels.

Another possible manufacturing method is the 'solid block' option in which the internal shape of the Y-connection is machined out of a solid steel ingot. This option is not considered to be suitable for the reasons given below:-

- A substantial increase in weight and therefore cost over the method proposed would result.
- Difficulties are encountered in achieving the requisite material properties in thick sections.
- Problems with machining a complex internal geometry in a large steel ingot may arise.

CONCLUSIONS

The technology required to produce a fully piggable subsea Y-connection to tie-in two gas pipelines exists and is now readily available.

The testing programme undertaken has resulted in a rational basis for the internal design of Y-connections. This includes a recognised method for predicting minimum flow velocities required for successful pigging. This flow prediction method together with the design principle established should prove useful in any future application of this technology to the design of Y-connections for a wide range of pipeline diameters and hydrocarbon products.

In order to make this 30-inch Y-connection economically attractive and to keep the weight within the limits of most manufacturers' capabilities, techniques have been developed which give significant savings in both weight and cost when compared to the 'solid block' options.

At the time of writing, tenders have been issued for manufacture of the Y-connection. The feedback gained from this and other testing procedures will allow direct comparison with the results of the structural analysis, and will lead to optimisation of the processes involved in development and designing piggable pipeline Y-connections.

MECHANICAL TESTING OF FLASH BUTT WELDED 36"OD API-X65 GRADE PIPES.

B.Lian & P.E.Kvaale.
Statoil - Norway.

1. INTRODUCTION.

The Flash Butt Welding (FBW) technique is developed by the Russians. They have used FBW for onshore pipelaying in the USSR for the last 20-30 years. The technique has for different reasons not been used in the Western part of the world. The main problem has been low toughness in the welded zone as well as uncertainty regarding soft heat affected zones in modern processed steels.

McDermott (McD) who have the license for use of the FBW technique in the Western world, have been looking for a project where the FBW technique could be used. Use of FBW would give a faster pipe laying speed, and economical benefits. A programme for development of the FBW technique was therefore started by Statoil and McDermott. This programme has consisted of two phases, the first was a study of X65 pipe materials behavior under simulated FBW thermal cycles. In the second phase, real flash butt welds have been tested. These welds were made with pipes delivered to the Statoil linepipe specification. Through the test programme, post weld heat treatment parameters have been developed by means of induction heating to a maximum temperature between 850 and 1000°C.

Volume 24: C. P. Ellinas (ed.), Advances in Subsea Pipeline Engineering and Technology, 229–248.

2. MATERIALS AND MECHANICAL TEST PROGRAMME.

36"OD x 1"WT pipes from four different steel mills have been used for testing of the FBW technique.

Traditional mechanical testing has been performed on FBW coupons made from all steels, while wide plate testing were only executed with welds from two steels.

The chemical composition of the steels used during these tests are given in Table 1, while Table 2 summarize the tensile properties.

Table 1 Chemical Composition.

Supplier	StA		StB		StC	StD	
Activity	A	B	A	B	A	A	B
%C	0.06	0.04	0.04	0.051	0.07	0.06	0.057
%Si	0.14	0.24	0.24	0.24	0.18	0.29	0.31
%Mn	1.42	1.3	1.16	1.13	1.51	1.34	1.34
%P	0.015	0.009	0.060	0.060	0.009	0.011	0.016
%S	0.002	0.0015	0.0004	0.0005	0.005	0.001	0.001
%Cu	0.01	0.02	--	--	0.01	--	0.30
%Ni	0.02	0.05	0.28	0.26	--	0.26	0.013
%Cr	0.04	0.05	0.002	0.002	0.03	0.02	0.013
%Mo	0.02	0.05	0.22	0.2	0.01	0.23	0.001
%V	0.03	0.0530	0.004	0.003	0.04	0.004	0.065
%Nb	0.037	0.042	0.034	0.034	0.037	0.037	0.043
%Ti	0.009	0.009	0.01	0.009	0.013	--	--
B(ppm)	2	3	--	--	1	--	--
Ca(ppm)	--	--	--	--	2	--	--
T.Al	0.036	0.027	0.014	0.015	0.029	0.024	0.032
CE	0.32	0.29	0.30	0.30	0.34	0.35	0.32
Pcm	0.14	0.13	0.13	0.13	0.16	0.16	0.16

A: Pipes for FBW trials (phase 2).

B: Material for simulation tests (phase 1).

Table 2 Tensile Properities.

Supplier	Transverse		Longitudinal	
	Yield MPa	Tensile MPa	Yield MPa	Tensile MPa
StA	467	546	483	564
StB	521	585	502	589
StC	497	544	484	561
StD	471	551	475	577

The test programme was not identical for all coupons. However, all welds have been tested with transverse tensile and side bend specimens. The Charpy-V test programme has as a minimum consisted of six specimens in the fusion line and three specimens one millimeter into the HAZ. For a few welds additional Charpy-V testing have been performed further into the HAZ. After completion of the Charpy-V test programme a retest was performed on some samples, which through the initial test programme showed excessive scatter in the Charpy-V results.

A CTOD testing programme in the fusion line, by means of Bx2B specimens were also performed on FBW coupons from all steels. The CTOD testing were performed on several welds with varying Charpy-V levels.

The hardness (HV5) indentations were positioned one millimeter below the outer and inner surface starting at the fusion line and measured till the base metal hardness was reached. In addition the extension and position of the soft zones were measured by hardness indentations for a few selected welds.

3. MECHANICAL TEST RESULTS.

3.1. Tensile Testing.

The results from the transverse tensile testing is given in Table 3.

Table 3. Tensile Testing.

Pipe Manufacturer	Weld No	R02 MPa	R05 MPa	R10 MPa	Rm MPa	Weld No	R02 MPa	R05 MPa	R10 MPa	Rm MPa
StA	125S	---	394	420	558	126S	---	387	413	553
	T	---	387	419	567	T	---	376	402	554
StA	127S	---	366	405	558	130S	---	360	394	561
	T	---	390	411	549	T	---	347	390	561
StA	134S	---	361	397	559	135S	---	388	409	547
	T	---	385	421	570	T	---	409	435	536
StA	136S	---	347	373	527					
	T	---	359	404	546					
StB	86T	441	479	---	550	89S	439	473	497	551
	P	439	479	---	553	T	431	459	480	544
StB	105B	482	514	534	564	108B	395	430	---	543
	S	420	457	484	536	S	418	452	---	553
StB	116B	419	455	---	535	117B	377	407	---	539
	S	412	450	---	547	S	384	412	---	528
StC	91T	439	475	500	575	95T	433	478	506	570
		392	425	438	562	96P	389	414	534	550
StC	121B	383	411	---	551	122B	378	410	---	554
	S	406	425	---	552	S	361	383	---	547
StC	123B	390	427	---	552					
	S	427	461	---	550					
StD	139S	---	426	467	574	140B	---	383	425	584
	T	---	411	457	592	S	---	397	438	588
StD	141B	---	371	408	532	142S	---	410	450	607
	S	---	391	425	536	T	---	422	462	598
StD	143S	---	443	479	607	144S	---	474	490	581
	T	---	442	484	602	T	---	429	458	569

3.2. Charpy-V Testing.

The Charpy-V testing has been performed in the fusion line with specimens positioned approximately 1 mm below the outer and inner surface. In addition testing has been performed in the "HAZ", both one and two millimeters from the fusion line. However, the main test position has been 1 mm into the HAZ, as the introductory tests showed that the toughness in the HAZ region where excellent for all steels. All test results is given in Table 4.

Table 4 Charpy-V test results /-30°C.

Steel	Weld No.	Upset Displ (mm)	Post Weld Temp °C (peak)	Hold time	Cooling rate Δt8/5	Charpy-V values Fusion line Cap	Root	FL+1mm Cap
StA	125T	8.2	-	90	-	140	132	213
			935	90	27	30	14	350
						48	70	352
	Retest					117	106	
						182	79	
						199	190	
StA	126S	10.2	780	120	38	83	62	353
			795	90	48	101	152	353
						99	22	284
	Retest						128	
							56	
							33	
StA	127T	8.6	905	120	38	128	144	180
			882	90	36	213	176	176
						44	171	300
	Retest					180		
						155		
						206		
StA	130T	7.0	838	120	30	150	136	288
			845	90	30	154	124	315
						138	53	353

Table 4 Charpy-V test results /-30°C. cont.

Steel	Weld No.	Upset Displ (mm)	Post Weld Temp °C (peak)	Hold time	Cooling rate Δt8/5	Charpy-V values Fusion line Cap	Charpy-V values Fusion line Root	FL+1mm Cap
StA	134T	7.5	928	120	31	128	327	113
			906	90	28	121	257	206
						174	208	192
StA	135T	10.5	-	120	-	113	84	353
			844	90	33	130	141	353
						168	40	353
	Retest						165	
							168	
							90	
StA	136T	7.5	920	120	30	225	112	353
			847	90	30	192	132	353
						138	186	353
StA	174S	9.2	1101	120	21	25	34	
			1080	90	20	257	19	
						30	40	
StA	175S	9.2	1049	120	22	27	16	
			1056	90	20	43	21	
						61	15	
								FL+2mm
StB	67B	9.0	897	120	35	142	130	354
			840	90	34	158	227	354
						44	99	219
								FL+1mm
StB	86T	8.9	937	120	31	168	95	353
			918	90	28	74	32	188
						28	71	214
StB	89T	9.3	960	120	32	175	211	354
			928	90	27	72	157	354
						161	67	354
StB	105T	8.8	873	120	32	207	34	213
			841	90	30	174	354	264
						189	354	354

Table 4 Charpy-V test results /-30°C. cont.

Steel	Weld No.	Upset Displ (mm)	Post Weld Temp °C (peak)	Hold time	Cooling rate Δt8/5	Charpy-V values Fusion line Cap	Charpy-V values Fusion line Root	Charpy-V values FL+1mm Cap
StB	108T	-	806	120	37	88	244	354
			808	90	33	118	85	354
						192	188	354
StB	108B					156		
						172		
						95		
StB	116T	7.6	873	120	33	51	354	354
			874	90	30	34	39	354
						139	145	354
StB	117T	8.9	796	120	40	354	55	354
			-	90	-	136	88	354
						228	27	354
								FL+2mm
StC	91S	10.2	823	120	48	28	65	262
			840	90	38	35	73	354
						19	71	110
								FL+1mm
StC	95S	-	1008	120	30	87	161	269
			934	90	29	101	79	354
						169	22	279
StC	96T	-	869	120	31	36	139	354
			776	90	39	65	73	140
						64	22	354
StC	121T	-	717	120	38	233	92	354
			798	90	45	139	75	233
						104	209	354
StC	121B					180	22	
						36	20	
						40	36	
StC	122T	6.8	867	120	30	133	209	197
			831	90	28	27	20	329
						354	10	354
StC	123T	5.9	763	120	48	168	140	353
			831	90	32	62	81	353
						33	71	353

Table 4 Charpy-V test results /-30°C. cont.

Steel	Weld No.	Upset Displ (mm)	Post Weld Temp °C (peak)	Hold time	Cooling rate $\Delta t8/5$	Charpy-V values Fusion line Cap	Charpy-V values Fusion line Root	FL+mm Cap
StD	139T	8.9	795	120	-	126	82	36
			859	90	36	33	78	163
						38	78	186
Retest						55		176
						87		342
						41		353
StD	140T	9.1	855	120	30	127	60	318
			-	90	-	42	166	87
						65	36	316
StD	140S					101	36	
						82	29	
						99	41	
StD	141T	9.1	900	120	34	145	111	189
			868	90	30	57	49	347
					-	92	39	354
Retest							60	
							58	
							82	
StD	142T	8.0	901	120	31	49	23	310
			877	90	30	29	33	353
						54	50	190
StD	143T	7.8	930	120	30	72	69	210
			920	90	30	64	38	215
						105	94	112
StD	144T	-	895	120	28	31	29	353
			912	90	28	111	33	353
						67	38	196

3.3. Side Bend Testing.

Two side bend specimens (t=10 mm) were made from all welds. These were bent over a 40 mm diameter mandrel (D=4t) to 180° with the fusion line in the center. No defects were observed during this test.

3.4. CTOD Testing.

CTOD testing were performed in the fusion line by means of Bx2B through thickness notched specimens. The test results are given in Table 5.

Table 5. CTOD test results at -10°C.

Weld No	Pipe	CTOD values
125B	StA	1.20(m);1.10(m) ----
126B	--"--	1.17(m);1.09(m);1.17(m)
127B	--"--	1.11(m);1.03(m);1.00(m)
130B	--"--	0.93(m);1.13(m);1.13(m)
134B	--"--	0.53(u);0.66(m);0.76(m)
135B	--"--	0.51(m);1.17(m);0.93(m)
136B	--"--	0.43(m);0.56(m);0.30(m)
86P	StB	1.08(u);0.65(u);0.98(m)
89B	--"--	0.85(m);1.09(m);0.28(u)
105S	--"--	0.37(u);0.51(u);0.67(u)
108S	--"--	0.37(u);0.23(u);0.23(u)
116S	--"--	1.17(m);0.46(u);0.65(m)
117S	--"--	0.78(m);1.07(m);1.16(m)
91T	StC	0.24(u);0.04(c);0.18(u)
121S	--"--	0.79(u);0.99(u);0.89(m)
122S	--"--	0.33(u);0.17(u);1.10(m)
123S	--"--	0.87(m);1.05(m);1.09(m)
139S	StD	0.53(m);0.46(m);0.24(m)
140T	--"--	0.27(u);0.39(u);0.71(m)
141T	--"--	0.23(u);0.43(u);0.40(m)
142B	--"--	0.41(u);0.81(u);0.27(u)
143B	--"--	0.18(u):0.32(u);0.34(u)
144B	--"--	0.31(u);0.34(m) ----

(m): max load
(u): unstable
(c): critical value.

3.5. Macro and Hardness Examination.

Macros were extracted from all welds. These have been examined for lack of fusion and possible gas porosities. In the fusion line small pores were observed in some of the macros. The maximum length and width of the pores were respectively 15 and 5 my. Photos in Fig. 1 show pores found in weld No.86. These pores were found randomly through the whole fusion line.

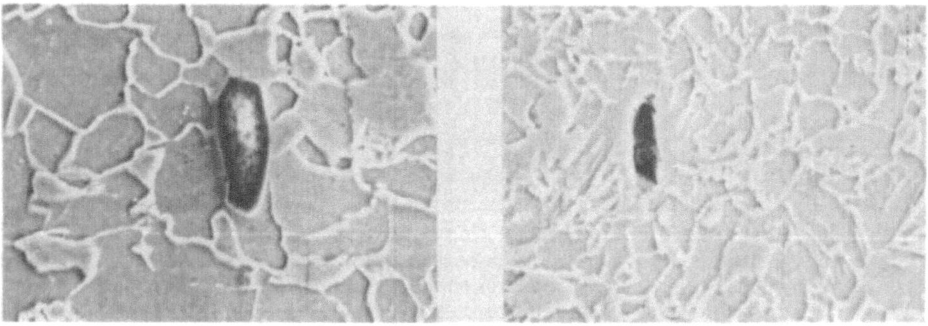

Fig.1 Pores in fusion line weld No. 86. 1500X.

The hardness were for all welds measured in the cap and root position from the fusion line till the base metal hardness were reached. To minimize the amount of testing the distance between each indentation varied between 2 and 8 mm. In Fig.2 a typical hardness distribution is given. For localization of the soft zones the hardness were also measured through the whole thickness on a few macro-specimens taken from the different steels. Fig. 3 show the result for one set of measurements. Hardness at different level is in that figure marked with different codes. In the white area no hardness below 180HV5 was found.

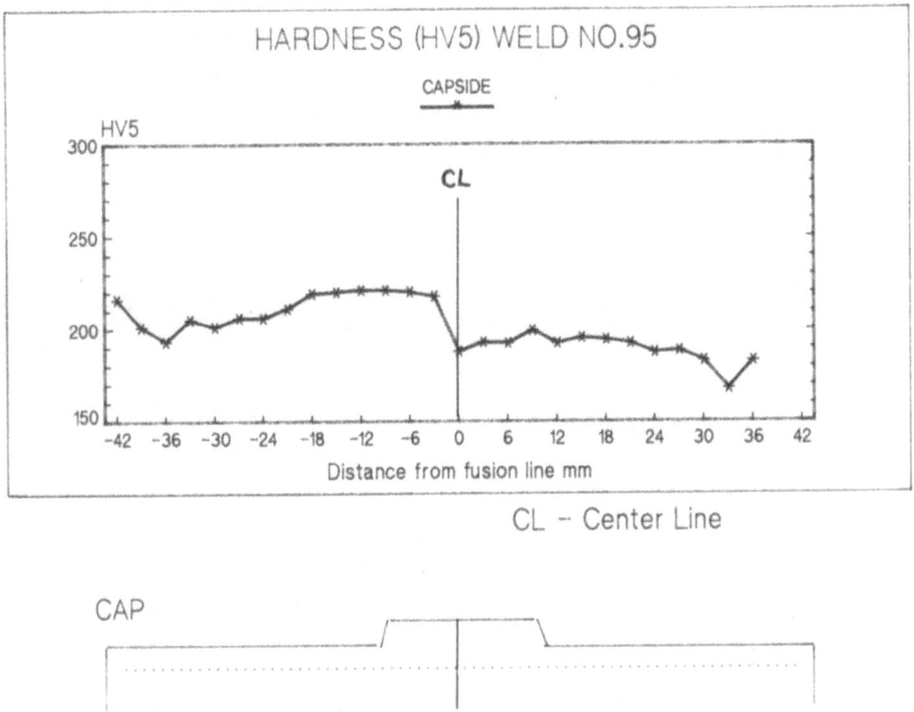

Fig.2 Hardness distribution in cap position for weld No. 95.

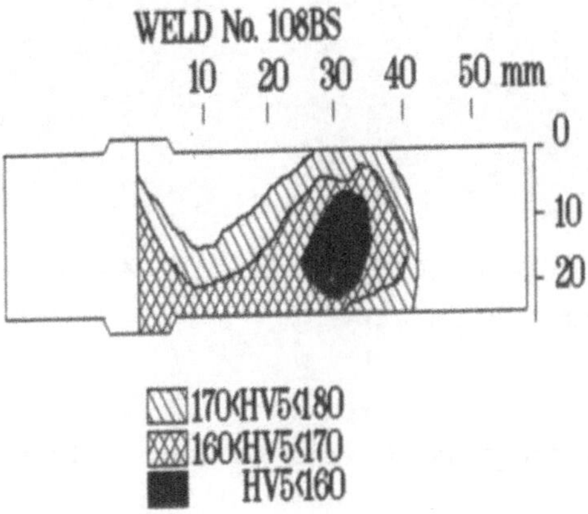

Fig.3 Hardness distribution (soft zones) through the plate thickness.

In Fig. 4 A-D the microstructure around the fusion line for weld No. 142 is shown. The grain size intercept is measured to 3.3my, corresponding to ASTM grain size No. 13. The microstructure is as expected ferritic with some carbide precipitates.

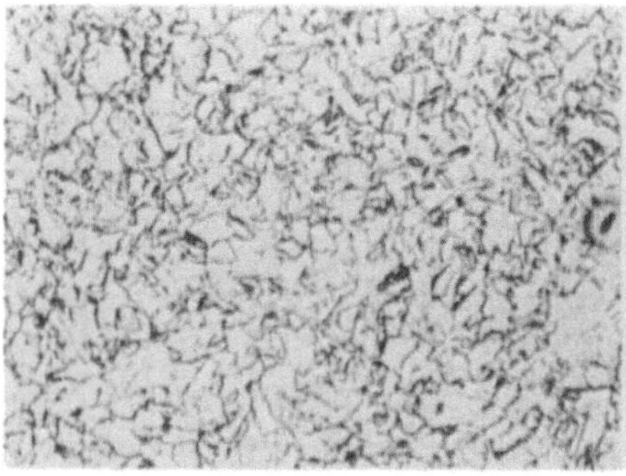

A. Microstructure at fusion line.

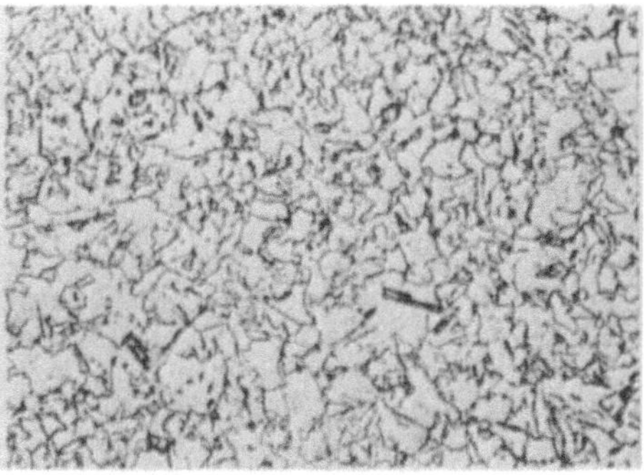

B. Microstructure in HAZ 2 mm from fusion line.

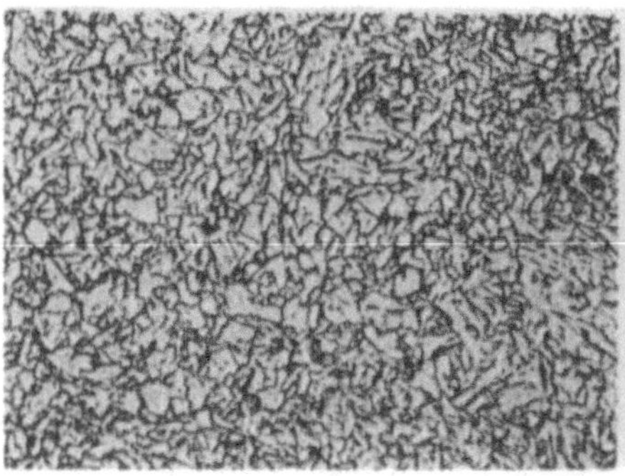

C. Microstructure in HAZ 3 mm from fusion line.

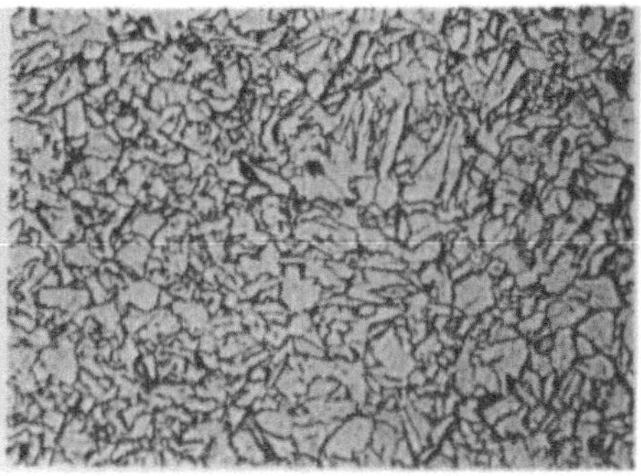

D. Microstructure in HAZ 5 mm from fusion line.

Fig.4A-D Microstructure in the HAZ for weld No. 142 pos. B 500 X

An examination of the base metal microstructure is also performed for all steels. In Fig. 5A-D the microstructure is shown. The microstructure is with exception for steel B, ferritic with a few islands of perlite. The microstructure in steel B is mainly bainitic, indicating a quenching process after completion of the plate rolling. The average grain size intercept for the base metals are, with exception of steel B, measured with results as shown in Table 6.

Table 6. Base metal grain sizes.

Steel	Grain size intercept (my)	ASTM No.
StA	5.3	11.8
StC	4.6	12.2
StD	6.8	11.1

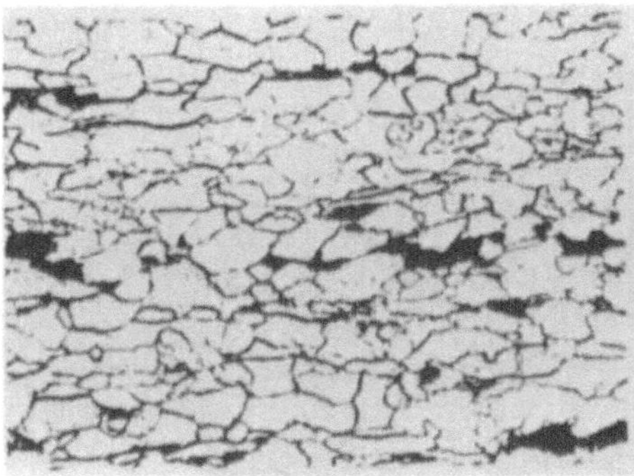

5A. Microstructure

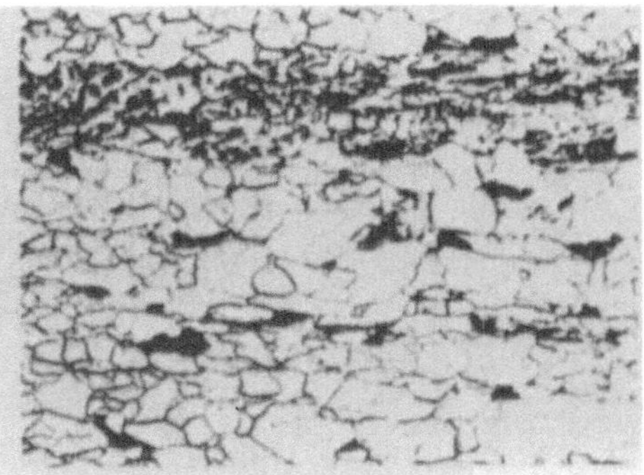

5B. Microstructure

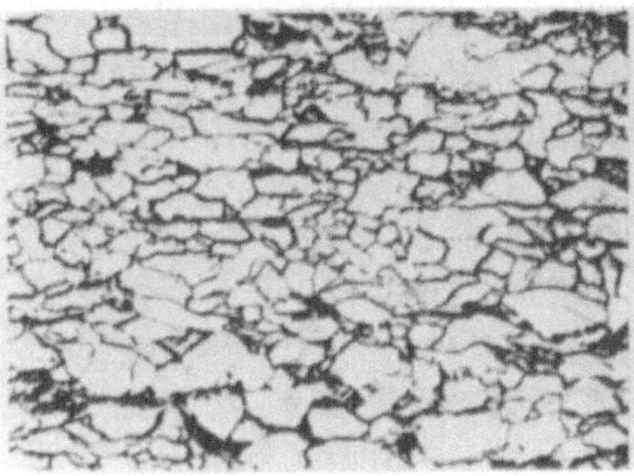

5C. Microstructure

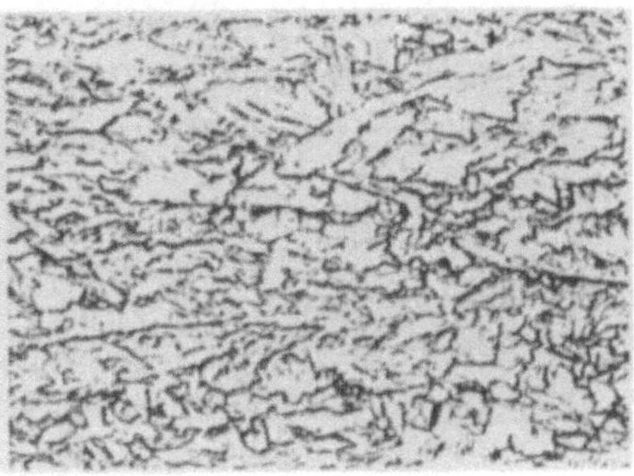

5D. Microstructure

Fig. 5A-D Microstructure in base metal 500X.

4 DISCUSSION.

4.1 Tensile Testing.

The tensile testing was performed by means of transverse tensile test bars 25 mm in width. For welds 86 to 123 an extensiometer was used to measure the yield point. For weld Nos. 125 to 144 the 0.5 and 1.0 % yield strength have been calculated from the test graphs. In Table 7 the average values and corresponding standard deviation (s) for the 0.5 % yield strength and tensile strength is given. These values are calculated from the results given in Table 3.

Table 7. Average tensile test properties.

Steel	R0.5%p (MPa)	s (MPa)	Rm (MPa)	s (MPa)	Rm/Rp
StA	375	19	553	12	0.67
StB	455	30	545	10	0.83
StC	431	31	548	18	0.78
StD	417	29	580	25	0.71

The difference in yield ratio may be explained by the test method. The extensiometer was used on the B and C steels while the yield strength was calculated from the tensile test graphs on steel A and D. Comparison of the standard deviations indicate that steel A is the most uniform steel after Flash Butt Welding.

However, as shown both in Table 3 and 7, the minimum tensile strength of 530 MPa (X65 quality) is satisfied for nearly all test specimens. The soft zones will therefore not cause any problems for use of the Flash Butt Welding technique in offshore pipelaying, provided that the PWHT is properly executed. That is also verified through the wide plate test programme.

4.2. Charpy-V Testing.

The Charpy-V testing has for all welds been performed in the fusion line both in "cap" and "root" positions. In addition testing has been performed 1 mm into the "HAZ" in the majority of cases. The results are given in Table 4. As seen, a wide scatter occurs in the fusion line, while the "HAZ" values in all cases are excellent. In Table 8 a summary of the test results are given for each steel.

Table 8 Summary of Charpy-V test results (Joules).

Steel	Fusion line				HAZ			
	Cap pos.		Root pos.		+1mm		+2mm	
	ave.	min	ave.	min	ave	min	ave.	min
StA	139	30	123	14	292	113	---	---
StB	140	28	146	27	324	188	309	---
StC	101	19	80	10	309	110	242	---
StD	74	29	59	23	246	87	---	---

The average values are for all steels acceptable, possibly with a question mark for the StD steel. However, the minimum values are not acceptable to normal specifications (34 Joule for Grade X65). The low values can in some cases be linked to fusion line porosity as shown in Fig.1, but in other cases no obvious reason is found. As the size of the pores in most cases are less than 5 my in width and length they may be present in specimens which visually occurs to be free for defects. However, a more reliable explanation to the low values may be found in the microstructure, where a ferritic band is found in the fusion line. Such microstructures are normally observed together with a low toughness.

4.3 Side Bend Testing.

Two side bend specimens were bent 180° over a 40 mm mandrel, without any significant defects. The pores which were observed during the macro examination did not cause any problems for these tests, which indicates that the material properties in the adjacent region to the fusion line is controlling the overall properties to the joint.

4.4 CTOD Testing.

A CTOD test programme was performed by means of Bx2B specimens with the notch through the fusion line. A summary of the test results is given in Table 9.

Table 9. Summary of CTOD test results.

Steel	Number of tests	CTOD values. Max.	Min.	Ave.
StA	20	1.20	0.30	0.89
StB	18	1.17	0.23	0.70
StC	12	1.09	0.04	0.65
StD	17	0.71	0.18	0.39

As shown, an acceptable CTOD level has been found for all steels. The low value (0.04) has later been related to a weld defect and oxide inclusions in the fusion line, located close to the notch tip. A chemical analyse of the inclusions showed manganese and silicon rich types. The low CTOD value is therefore not representative for the welding method as such. As the CTOD testing has given more uniform values than the Charpy tests, it is reasonable to believe that the "brittleness" causing the wide scatter in the Charpy-V results is not critical for the joints global performance. This assumption is based on the fact that a CTOD test in full wall thickness normally is more stringent than a Charpy-V test. However, before any final conclusion is drawn a comparison with the wide plate test results should be performed.

4.5. Macro and Hardness Examination.

In some of the macrospecimens microporosity were found in the fusion line, as shown in Fig.1. However, no obvious reason to the pores were found. The actual upset length varied during welding, but no correlation to porosity could be documented. The upset speed was also discussed as a possible reason for the pores. To test that hypothesis two welds (Nos. 174 & 175) were made with about 20 mm in wall thickness. In these welds no porosity were observed, but as seen in Table 4 a poor toughness level was found. Once again, no obvious explanation is found. The peak temperature during PWHT of these welds may be to high, close to 1100°C. However, additional tests have to be made before any conclusion is drawn.

Examination of the microstructure in the fusion line area is shown in Fig. 4A-D. As seen, a very fine grained ferritic structure without any bainitic islands occurs. This is also confirmed by the hardness tests which showed no hardness above 300 HV5 for welds in full wall thickness, while it on weld Nos. 174 and 175 with reduced wall thickness, hardnesses above 350 HV5 were found. That may also explain the reduction in toughness which occurred in the latter welds. As the microstructure in the fusion line area is extremely fine grained, no further improvement in mechanical properties seems possible through grain refinement. However, it is still possible that the repeatability of the PWHT programme can be improved, and as such a more uniform Charpy-V level obtainable.

4.6 Wide Plate Tests.

The wide plate results can shortly be summarized as follows.

The wide plate tensile tests done at room temperature on 200 and 300 mm wide panels indicate that the HAZ softening caused by the weld thermal cycles is not a matter of major concern, provided a post weld normalizing treatment applied at a sufficiently high temperature is followed by a proper cooling programme.

The wide plate tests on defective flash butt welds have shown that small natural defects, originating from an improper weld cycle, still produce acceptable weld joint performance in as far the fracture initiation resistance is concerned.

The wide plate tests done at -10°C on pipe sections, containing a precracked surface crack, located at the inner side at the flash butt weld interface gave very promising results. Surface cracks of 98 mm in length by a maximum of 6 mm in depth did only induce unstable fracture after significant plastic straining. It should be emphasized, however , that additional wide plate test data are a prerequisite before allowable crack sizes can be quantified with more confidence.

These results documents our findings during the CTOD test programme. After the weld normalizing process, plastic strain is necessary to obtain an unstable fracture in both cases. The low Charpy-V values which randomly occurs in the fusion line is therefore of no significant concern regarding the joint overall fracture initiation properties. However, the reason for the low Charpy-V values should be further investigated.

5 CONCLUSIONS.

The minimum required tensile strength for X65 steels was satisfied for nearly all tensile test bars extracted from welds which have been heat treated properly after flash butt welding. These results were also documented by the wide plate test programme. It is therefore concluded that the anticipated soft zones after flash butt welding is of no major concern.

The Charpy-V testing gave a wide scatter in absorbed energy from specimen to specimen. It is not clear if this is related to the test method, unpropper heat treatment or minor weld defects. Further investigations should therefore be performed before any conclusion is drawn. However, as the welding method easily will give a ferritic band, less than 100 my in width, in the fusion line a scatter in Charpy-V values may be expected. It should therefore also be analysed if the Charpy-V testing method is suited for flash butt welding.

The side bend testing proved that the toughness in the weld zone were sufficient to avoid initiations of cracks in the fusion line. This result does also indicate that the low Charpy-V values which occurs occasionally do not represent the joints overall properties.

The CTOD test programme gave surprisingly uniform values compared to the Charpy-V results. The CTOD level was high with average values above 0.4 mm. These values documents that a significant plastic strain is necessary before unstable fracture occurs.

The macro examination revealed some micropores less than 15 my in length in the fusion line. The overall testing has however shown that these pores have no significant influence on the joints mechanical properties. The microstructure examination documented an extremely fine grain ferritic microstructure in the heat affected zone. As the joints are cooled by a water quenching after heat treatment there is a risk for transformation to a martensitic microstructure. These examinations have however not shown any martensitic islands in the heat affected zone. That may however be the case for weld Nos. 174 and 175 which were made on pipes with a reduced wall thickness.

The hardness examination confirmed the microstructure findings as no hardness above 300 HV5 was found for welds with full plate thickness. The maximum hardness in weld Nos. 174 and 175 were however above 350 HV5 in some spots, which indicate a martensitic structure in these parts. That may also explain the poor test results observed for these welds.

The wide plate test results gave all acceptable results. These results were in agreement with the CTOD results and as such they do also indicate that the scatter in Charpy-V values is not representative for the joints overall properties.

THE CENTRAL BRAE PROJECT RETROFIT RISER INSTALLATION

R. FARROW AND N. CRESSWELL

Marathon Oil U.K., Ltd.

PROJECT SUMMARY

The Brae field, discovered in 1975, is located in block 16/7 about 42 nautical miles north west of the Occidental Piper field and 70 nautical miles of the BP Forties field. The Brae 'A' platform was installed in 1982. First oil was exported in July 1983 and production rates of around 100,000 barrels per day have been achieved.

The Brae 'B' platform was installed in 1987 and first oil was achieved in April 1988. The Brae 'A' and Brae 'B' platforms are connected to each other by a twin 18" pipeline system. from the Brae 'B' platform, oil is transported to Brae 'A' where it is combined with the processed Brae 'A' oil. The combined oil is then transported via the 30" submarine pipeline to BP's Forties Charlie platform where it enters into BP's 32" submarine facility at Kinneil.

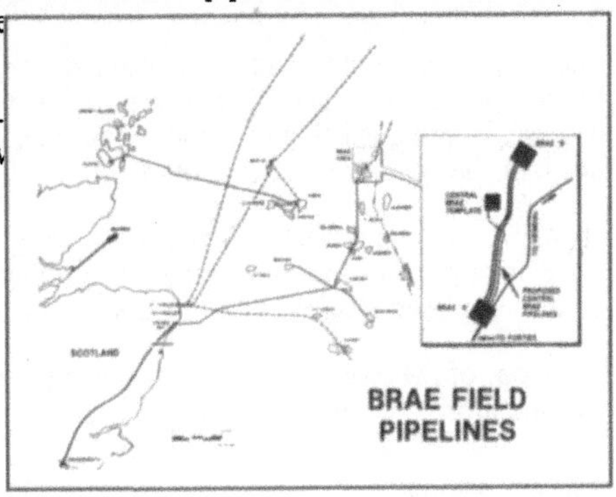

BRAE FIELD PIPELINES

The Central Brae project is Marathon's first sub- dedicated team of engineers and support staff draw A major part of the project are the 6" production, 4" service, and 10" water injection pipelines, the retrofit caisson riser and the chemical and control umbilicals. Together these represent over a third of the total development cost.

The development plan for Central Brae was approved by the Brae group participants in October 1987, and Department of Energy approval was received in April 1988.

This paper presents a general description of the project together with a more detailed account of the design, fabrication and installation of the retrofit caisson bundled riser system.

Volume 24: C. P. Ellinas (ed.), Advances in Subsea Pipeline Engineering and Technology, 249–261.

Extent of Project

The Central Brae reservoir, discovered in 1976, lies between the Brae 'A' and 'B' platforms, and is divided into two regions. The proven area, defined by four successful wells drilled to date, and a 'possible' area, characterised by heavy faulting and one suspect well. The proven region is estimated to contain 162 mmstb of oil in place whilst the 'possible' area has estimated oil in place of 111 mmstb. The estimated recovery from the proven area using two producers and two water injectors to supplement some natural water influx is 64 mmstb of oil. The estimated peak production rate is 20,000 bbls/d.

The artist view below illustrates the main elements of the project.

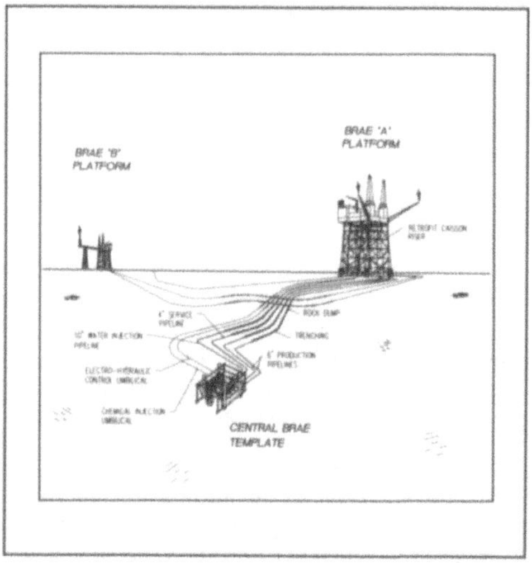

ARTIST VIEW

A template was tied back to the Brae 'A' platform with 2 x 6" production, 4" service, and 10" water injection pipelines. There are also two umbilicals, one for chemical injection and the other for controls.

New risers were required on the Brae 'A' platform for the pipelines and an existing J-tube was used for the umbilicals.

All the pipelines were trenched, except where they cross over the existing Brae 'A' to 'B' 18" pipelines. The production pipelines are also covered with 1m of rock to prevent upheaval buckling.

To avoid hydrate formation and the cooling effect that the cold Central Brae oil would have on the Brae 'A' processing facility, thermal insulation of the production pipelines was required.

The template has 8-slots and is installed over two existing delineation wells which will be reclaimed. Initially production will be from one of these and one new on-template well, whilst water injection will use the other existing well and a new one. A fifth delineation well is to be drilled after start-up, into the 'possible' area of the field.

To allow maximum flexibility the template is designed to allow any well to be used for either injection or production or for change over during the life of the field. The subsea wells are controlled by a multiplexed electro-hydraulic control system via the control umbilical.

The scope of the project also includes the installation, on the platform of the topside piping, control and monitoring equipment, and the chemical injection and well kill equipment.

Pipeline Routing

A plan of the pipeline route is shown overleaf. The existing Heimdal and Brae 'B' pipelines form a narrow corridor in which the Central Brae pipelines were laid. The pipelines exit the corridor by crossing over the existing trenched 18" lines. The final route selection was essentially a trade-off between the conflicting requirements to:-

- keep within the narrow corridor.
- space out the pipelines as much as possible, so as not to make pipelay positioning difficult, and allow trenching as close as possible to the platform.

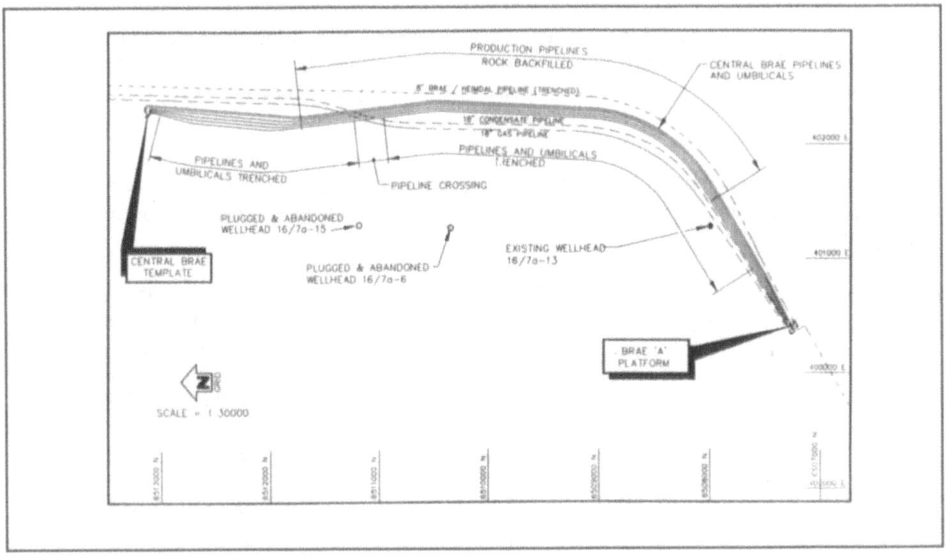

At the platform end the design also had to:-

- ensure the expansion loops did not block access to the J-tube.
- maintain a corridor for future pipeline and power cables to access the remaining J-tubes.

Dual lay of either the 6" duplex or the two umbilicals was evaluated as a way of increasing separation between the lines. However, it was felt that the disadvantages of the more difficult construction methods involved out-weighed the benefits gained, therefore these options were not progressed further.

The narrow corridor in which the lines were to be laid meant that, by design, the lay accuracy had to be within + or - 5m. Experience with other pipelines had shown that this degree of accuracy could only be achieved by using sub-sea acoustic positioning for the entire pipeline length. This method was therefore specified as the primary positioning system.

Existing Facilities

The Brae 'A' platform does not have sufficient spare risers or J-tubes to accommodate all of the Central Brae pipeline and umbilical needs. There are spare risers on the platform but none of these are of the correct size. There were also three 18" J-tubes which were not in use. However, one of these was already allocated for a future incoming pipeline and the other allocated for future power cables.

The Brae 'A' jacket is battered on the east and west faces but is vertical on the north and south faces.

CAISSON RISER DESIGN

Riser Options

Studies were commissioned to evaluate the various options for pipeline risers. A conductor access system was developed whereby rigid risers would be installed in a spare conductor slot. Whilst technically feasible, this was not selected due to the more difficult subsea tie-ins and the amount of drilling mud which would need to be moved from the seabed.

The merits of both individual risers and caisson bundled risers, in various locations were reviewed in some detail. The factors which influenced the comparisons included; the extent of clamp installation and seabed preparatory work required, avoidance of supply boat loading areas, maintaining access to un-used J-tubes, proximity to existing pipelines and flare booms, ability of the jacket to take the weight, and use of a vertical or battered jacket face.

After reviewing various permutations, the option selected was for installing, external to the platform jacket, a single caisson bundled riser for all pipeline risers and using a single J-tube for the support of the two umbilicals.

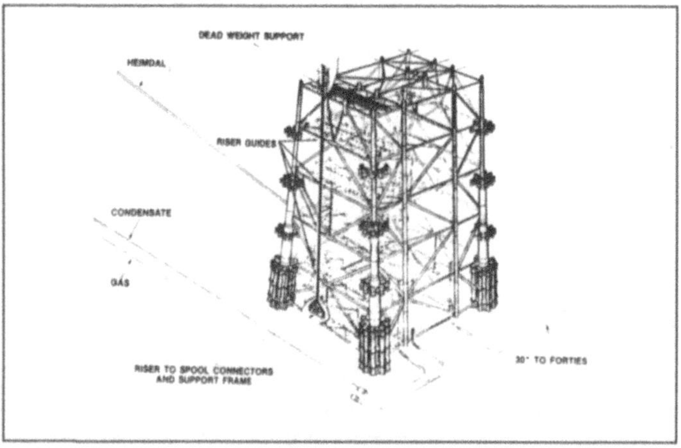

When compared to other systems the use of a bundled riser provided a means of installing all of the Central Brae riser needs with a single installation lift, thus reducing installation vessel time and costs.

The north face location was selected as the most appropriate. The main factors which influenced the location were the avoidance of both the blockage of access to the un-used J-tubes by the pipeline's expansion spools and the avoidance of the supply boat loading area.

This vertical platform face also made lifting into the dead weight support and sub-sea clamps easier. The location was however, between the two large flare booms and the proximity of the crane barge's jibs to these had to be carefully evaluated, to ensure that access could be achieved.

Design Philosophy

The design philosophy was for each riser to be installed and freely suspended within a 34" caisson, expansion would be allowed by movement of the individual riser pipes through sliding seals in the base plate. The design allowed for the riser bundle to be fabricated and assembled before insertion into the caisson thereby allowing inspection on the completed items (e.g coating).

The entire caisson riser was suspended from a dead weight support welded to the platform at the EL +14 m level. To minimise installation time the design of the caisson support had to be such that the transfer of the caisson riser from the lift barge to the platform dead weight support could be completed as quickly as possible.

A system was devised whereby horizontal trunnions on the caisson could be set down onto the lower halves of the two dead-weight support clamps and then released from the lift barge crane. Final closure of the sub-sea and dead weight support clamps could then be made after the lift barge had departed.

The design incorporated a guidance system to guide the caisson, both vertically and horizontally into the cups of the dead weight support.

Caisson Contents

Caisson sizes of between 30"- 36" were reviewed for their ability to accommodate the project's immediate and future requirements. A small increase in the caisson diameter significantly increases its ability to accept additional risers. It was decided to include two spare production risers in the caisson to be consistent with two spare production headers included on the template. For this number of risers a 34" caisson was, therefore, required.

A cross-section of the 34" caisson is shown below. In addition to the production, service, and water injection lines, the caisson also includes instrument lines. One instrument line contains a bundle of small bore hydraulic pipe for use with any subsea safety valves that may be fitted at a later date.

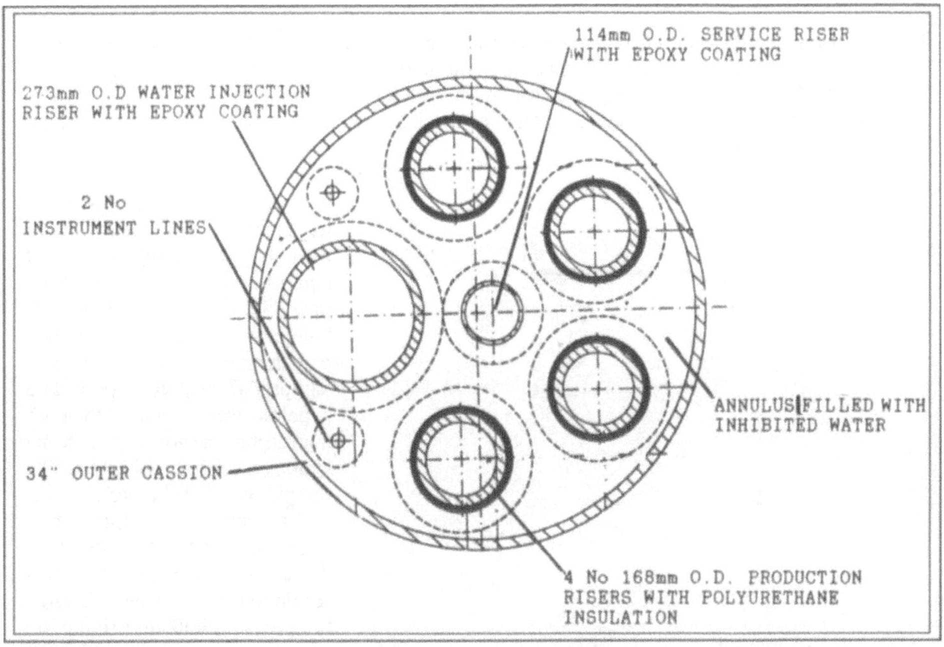

Materials

MATERIAL SUMMARY						
	PRODUCTION		INJECTION		SERVICE	
	PIPELINE	SPOOL	PIPELINE	SPOOL	PIPELINE	SPOOL
MATERIAL TYPE	DUPLEX	INCONEL/ CLAD C/S	C/STEEL	C/STEEL	C/STEEL	C/STEEL
GRADE	UNS 31803	INCONEL/ 5L x 52	API 5L x 65	API 5L x 65	API 5L x 52	API 5L x 52
DIA	163 O.D.	170 O.D.	273 O.D.	273 O.D.	114 O.D.	114 O.D.
DESIGN PRESS	3600 PSI	3600 PSI	5000 PSI	5000 PSI	5000 PSI	5000 PSI
DESIGN FLOWRATE	20,000 BBL/D	20,000 BBL/D	70,000 BBL/D	70,000 BBL/D	6 mm SCFD	6 mm SCFD

The table above summarises the materials used for the pipelines.

The Central Brae oil contains hydrogen sulphide and carbon dioxide and has a wellhead design temperature of 111 degrees centigrade. It is therefore very corrosive to normal carbon steels. Duplex stainless steel had to be used. However, for the template and expansion spools internally clad carbon steel was used. This was only available in short 3-4m lengths and was not therefore suitable for the pipelines.

To confirm the satisfactory resistance of welded duplex pipes to pitting, crevice corrosion and stress corrosion cracking a corrosion test programme was carried out. The tests were performed at temperatures and pressures which simulated the service conditions expected.

A further test programme was carried out to assess the effects of cold work, induced by the reel pipelay method, on the duplex.

These test programmes confirmed satisfactory performance for our service conditions.

Pipeline hydraulic simulations had determined that in order to achieve the design arrival temperature of 32°c at Brae A at a flow rate of 5000 bpd, the maximum heat loss from the pipelines was 4 w/m²/°c. To achieve this a pipeline insulation system was required. A review of available insulation systems and materials revealed potential unsuitability for either reel laying or to withstand the high operating temperatures.

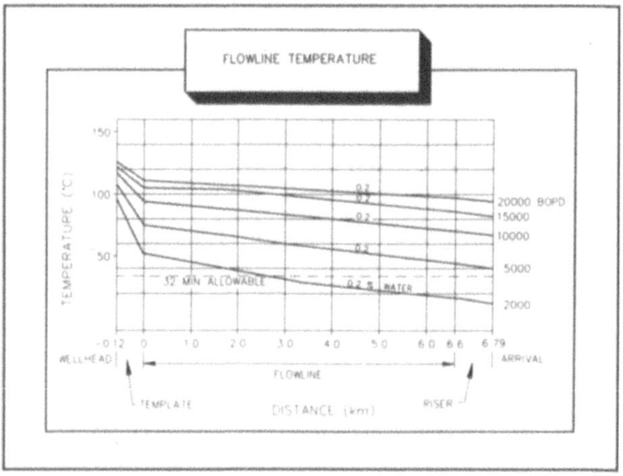

Conventional elastomer systems (e.g Neoprene or E.P.D.M.) were thought to have the most potential, therefore to evaluate their ability to withstand the temperatures involved a test programme was commissioned. Samples of neoprene, epdm, and hypalon were applied to both duplex and carbon steel pipe by various specialist applicators. The samples were left immersed, with C.P potentials applied, in a cold water tank and hot oil of 115°c (70°c for neoprene only) was continuously circulated through the pipe to simulate production conditions. Additionally, one sample of polyurethane field joint coating material was applied to duplex for testing.

Inspection of the test samples after test periods of 3 and 6 months revealed varying amounts of disbondment in all of the polychloroprene on duplex pipe samples. However, the polyurethane sample still had excellent adhesion. This material was therefore selected for both the pipelines and risers.

For both the water injection and the service pipelines conventional carbon steel materials were selected with fusion bonded corrosion coating specified for both the main pipe and the field joints.

Design Features

The size of the expansion spools at the base of the risers were constrained by the close proximity of the existing 18" pipelines which were only 22m away. To avoid adding additional expansion spools elsewhere along the pipeline the lower termination arrangement was evaluated to determine the amount of movement it could tolerate without over stressing the pipe or the flanges fouling each other. It was found that up to 1m of pipeline expansion could be allowed and this was sufficient to avoid additional spools.

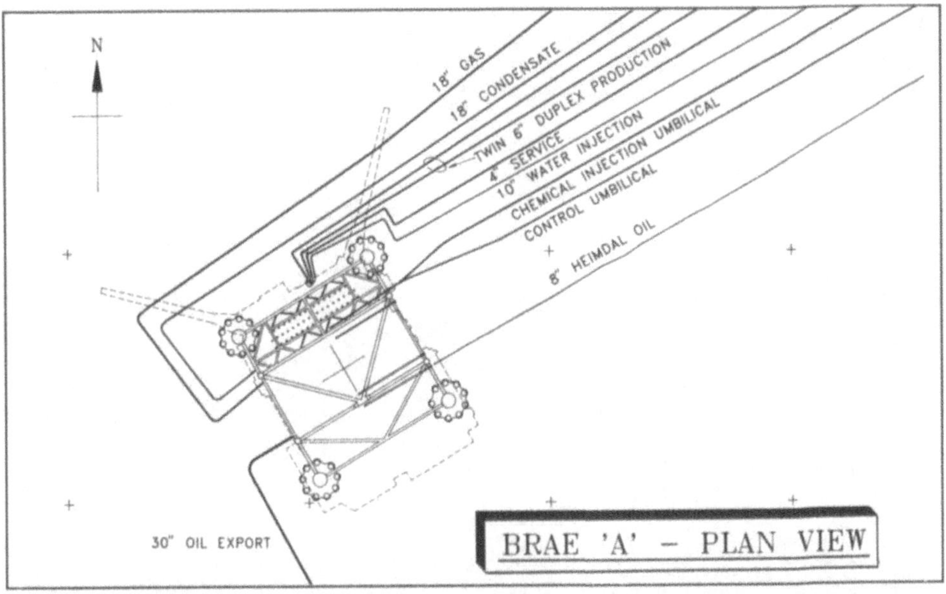

The subsea clamps were of the conventional box type construction except that the lowest clamp had an additional fork type guide to funnel the caisson into the clamp.

A protective cage was fitted around both the top and bottom termination to avoid damage to the risers during installation. (See photos on Page 8).

ONSHORE FABRICATION

Fabrication and Assembly

The task of translating the design into reality commenced in June 1988 with the award of a fabrication contract.

At the fabrication yard much thought was given to the site layout of the caisson because its length could have blocked access. As mentioned, the construction philosophy relied on the completion of the internal riser bundle before pulling on the caisson. The key to this procedure lay in the design of the internal baffles that were required to maintain separation of the risers.

Each of the 16 cast nylon baffles, which were fitted every 8m, comprised of 3 separate sections. The sections allowed the outermost risers to be fabricated outside of the bundle before being rolled into the open holes of the baffles. This allowed access for welding and coating to be maintained. The duplex riser pipes were delivered to site from a specialist coating works complete with polyurethane coating. On site coating was therefore limited to the individual field joints which were coated using moulds.

When all the risers were complete the baffles were closed and bolted. The entire assembly was then inspected before adding the outer caisson. (See photos on Page 9).

Although friction alone would have maintained the baffles in position, as a precaution, small clamps were secured on to the central 4'' service line either side of each baffle to ensure they would not drop if friction was lost.

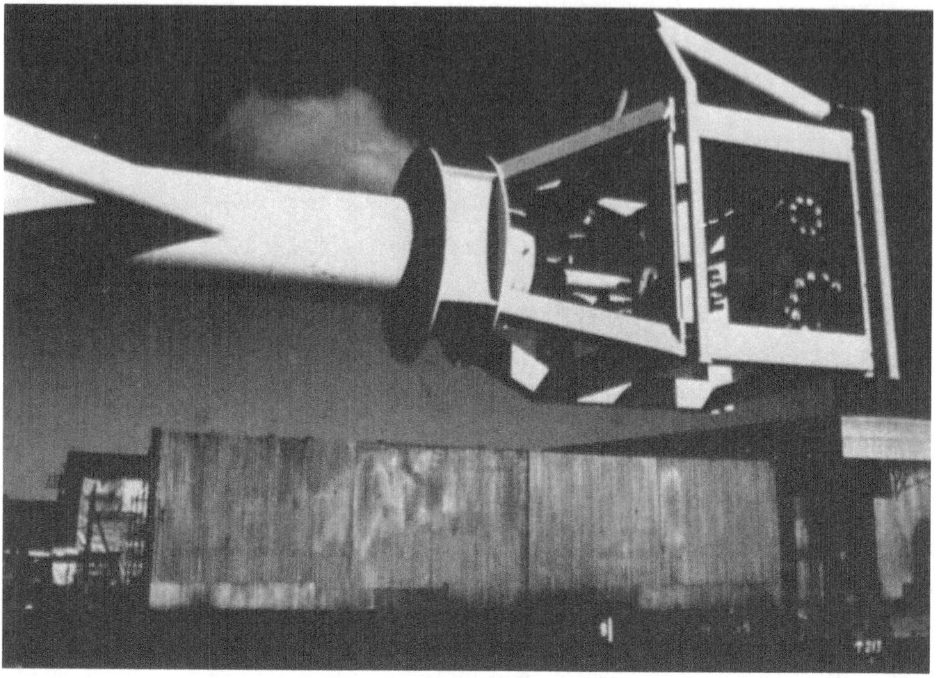

RISER TOP PROTECTIVE CAGE

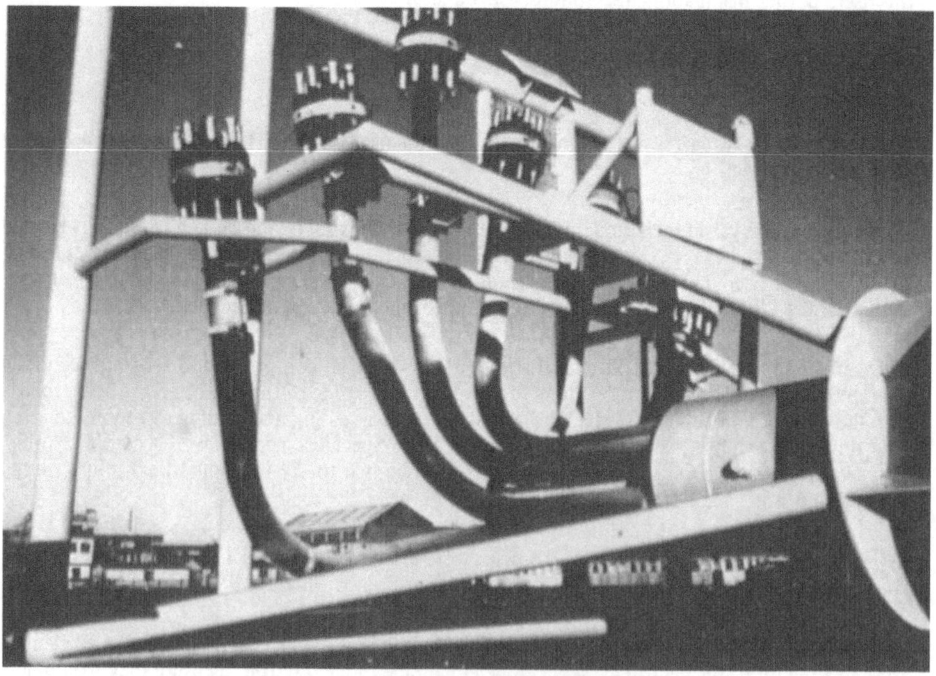

RISER LOWER TERMINATION

RISER BUNDLE ASSEMBLY

RISER BUNDLE BEING INSERTED INTO CAISSON

Each baffle had small wheels fitted in it's outer edge which allowed the caisson sections to ride smoothly over the riser bundle during assembly. Each section of the caisson was welded to the next just beyond the end of the bundle to eliminate the risk of damage to the riser coatings. Once all sections of caisson were installed, the end termination flanges were fitted. The top and bottom spools and protection frames were then fitted and the risers' hydrotested finally.

The internal space within the caisson was left void for filling with inhibited water offshore and the risers hydrotest water was emptied to minimise weight.

Strong Back

From an early stage in design it had always been Marathon's preference that the 132m long caisson be supported during rotation from the horizontal to the vertical with a supporting pipe. Although analysis had shown that an unsupported caisson could survive a four point lift, the more cautious supported approach was adopted. A 110m long by 84" diameter pipe, called a strongback, fitted with a series of hinged clamps to support the caisson, was used for this purpose.

Load Out

To accommodate the caisson's length a 400' x 120' cargo barge was selected for transportation. Even so, the caisson overhung both the bow and stern by 5m. The strongback had already been installed on to a barge prior to delivery to the fabrication yard. It was only necessary therefore, to lift the caisson riser into the clamps using 3 crawler cranes. Before securing it for the voyage it had to be turned over 180°. This was achieved by resting the caisson on strips of PTFE already set in the clamps and turning it.

The caisson riser's flexibility was evident during this load-out, as the weight of the upper and lower terminations produced a deflection of at least 600mm at each end.

LOAD OUT OF RISER

OFFSHORE INSTALLATION

Installation Engineering

Considerable engineering effort was provided by the installation contractor. Although this was mainly concerned with the lifting analysis, seafastenings, rigging arrangements, etc, significant input into the caisson design, especially the guidance system on the dead weight support, and lifting points was also provided.

On the platform two flare booms project 120m from each corner which presented a significant obstacle to the crane vessel, particularly when flaring gas. Procedures were developed to overcome the restrictions the flares imposed and to establish contingencies in the event that major flaring occurred.

Preparatory Work

The dead weight support and a bumper rail, to protect topside pipework from the lifting slings, were welded to the platform prior to mobilisation of the crane barge to the field.

Additionally, the installation rigging was installed, and a lifeboat station and a stairway down to the +8m walkway level were removed to allow the caisson to be positioned.

A diving vessel was used to pre-install the subsea clamps on to the horizontal jacket members and to remove drilling mud from the seabed at the base of the riser.

Both the platform and diving work was hampered by the bad weather which occurred at this time. In the case of the deadweight support the scaffolding was "re-arranged" by waves on at least two occasions and had to be rebuilt. Significant time was also lost on diving work due to weather.

It was known that drilling mud deposits on the seabed existed in the area of the caisson. The amount of material was found, by a detailed ROV grid survey, to be more extensive than previous surveys had indicated.

Excavations to remove this material were performed using the RUE tool which dispersed it with a high volume stream of water. The tool was cross hauled along the sea-bed towards the platform and by making a number of parallel excursions the whole area could be covered. After several days work, divers confirmed successful removal of a large volume of material. However, there were difficulties in cross-hauling when close to the platform and during the period considerable weather delays were experienced. Although sufficient material was removed to install the caisson, at the time of the crane vessel mobilisation only 30% of the total scope had been completed and the remaining scope had to be re-scheduled.

Transportation

Both the template and the riser installations required a crane vessel and supporting transportation barge. Since different fabrication yards were used, the caisson riser was first loaded out onto the barge which was then towed to the second fabrication yard for load out of the template, before mobilisation to the Brae field. By having both items available in the field together, maximum flexibility of installation sequence was maintained up to the last minute to suit prevailing weather conditions and platform operational needs.

Rotation and Installation

On April 12th, the heavy lift vessel, anchored near to the Brae 'A' platform, lifted the strongback and caisson riser from the transportation barge and performed the rotation to the vertical using its 2 cranes. At the top of the strongback a large clamp, designed to take the entire caisson weight once vertical, was kept bolted, all other clamps having been released. The whole assembly was then hung off the stern of the crane vessel with two 6" dia slings, before the crane lifted the caisson out of the strongback's top clamp, ready for installation on to the platform.

These operations needed a total of 8 lifting slings and 6 other secondary positioning wires. All of these festooned the strongback and care was taken by the yard in tying down each sling in reverse order of its removal.

Before the crane vessel moved closer to the platform, the OIM had isolated the upwind flare. Previous studies had shown that in the event of a platform shutdown, the heat from only one flare would not be sufficient to damage the crane barge's steelwork before action could be taken. Once the crane vessel had set up in its final position, within 30m of the platform, rigging lines designed to pull the caisson under the cantilevered walkways above were connected, and the pull-in started. The guides on the dead-weight support worked well and funnelled the trunnions into the support clamps. (See photos overleaf).

The rotation and hang-off of the strongback and subsequent separation of the caisson, were the most weather sensitive operations and took 7 hrs to complete, whilst the actual installation another 3 hrs.

Once in position the sub-sea and dead weight support clamps could be closed and tightened. This occurred after the lift barge had departed. The guidance forks and vertical posts on the dead weight support were removed.

OVERALL SCHEDULE

The project offshore schedule is shown on the bar chart below. Offshore construction commenced on March 1st with diving preparatory work for the caisson riser installation and was completed on 4th July prior to arrival in the field of the drilling vessel. First oil was achieved on September 1st 1989.

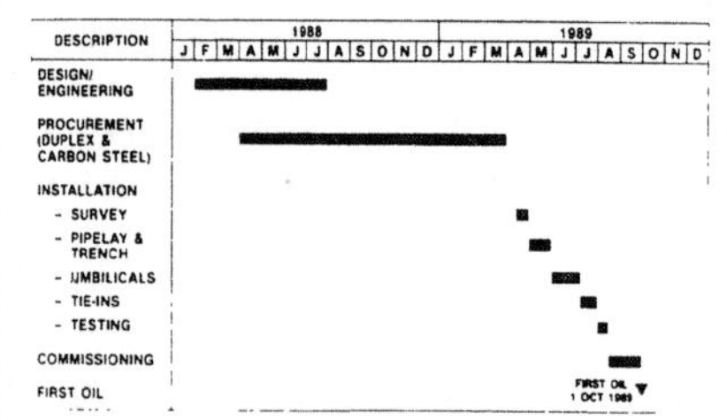

Central Brae Pipelines Schedule

THE BRAE GROUP

Marathon are operators of the Brae field on behalf of the Brae group participants who include:-

Marathon Oil (UK) Ltd (operator) 38%
Britoil plc 20%
Bow Valley Exploration (UK) Ltd. 14%
Kerr-McGee Oil (UK) plc 8%
Gas Council (Exploration) Ltd. 7.7%
LL&E (UK) Inc 6.3%
Sovereign Oil and Gas Ltd......................4%
Norsk Hydro Oil and Gas Ltd. 2%

RISER LOWERING INTO DEAD WEIGHT SUPPORT

RISER LOWERING ADJACENT TO BRAE 'A'

HAZARD ASSESSMENT AND PIPELINE DESIGN

Dr K A J Williams

AME Ltd, Surrey Research Park, Guildford

ABSTRACT

As a result of recent developments it is expected that Formal Safety Assessments will have to be performed for all offshore installations in the UK Sector of the North Sea. In this paper the main elements of these assessments and application to pipeline design are briefly introduced. The information required for these assessments is discussed and sources of historical data and information that are available are presented. Finally incidents which have caused damage to pipelines in the past are described.

INTRODUCTION

The design of a system is regulated in order to ensure that it can perform its expected function safely. This is usually achieved by following appropriate Codes and Standards which have been formulated over the years on the basis of experience. However, for some time there has been concern that reliance on good engineering practice, the application of approved standards and certification and inspection are not by themselves always sufficient to identify and highlight

Volume 24: C. P. Ellinas (ed.), Advances in Subsea Pipeline Engineering and Technology, 263–278.

hazards and sequences of events that can lead to a major hazard [1], and thus ensure that required levels of safety are achieved and maintained.

Following the issue of a discussion document by the UK Department of Energy [1], it is anticipated that Formal Safety Assessments will have to be performed for all offshore installations in the UK Sector of the North Sea. The perceived objective of these studies is to confirm that all possible hazards have been recognised and steps taken to prevent or minimise their occurrence and the consequences of such an occurrence.

Hazards to be considered for an installation include product release from pipeline systems and therefore the safety assessments will need to include hazards to pipelines and their consequences. The analysis techniques that it is suggested should be used in these assessments include Quantified Risk Analyses, Detailed Safety Evaluation Studies, Evacuation Studies, Ship Collision Studies and Safety Audits. The main elements of these assessments will be the identification of potential hazards, the analysis of their consequences and the determination of their expected frequency. In this paper these steps and their use in pipeline design are briefly discussed.

Three main areas of difficulty in the quantification of risk have been identified by the Department of Energy [1]. These are the limited statistical data on equipment failure rates on which to predict overall failure criteria, the level at which acceptability criteria for failure rates can be set and the assessment of human factors and their contribution to such risks. The information that would be required to overcome these difficulties for pipelines is discussed and the sources of historical data and information that are available are presented.

USE OF SAFETY ASSESSMENT IN PIPELINE DESIGN

The use of methods such as HAZOP and FMECA and reliability analysis techniques such as Fault and Event Trees is well established. However, they have only been used to a limited extent in pipeline design. During the design process hazard and risk analysis techniques have been used mostly as a decision making tool to assist in the choice between design solutions. An example of an area where this type of analysis has been used is pipeline routeing, in accordance to guidance given in IP6. The IP6 supplement [2] states that routeing should take into account access and safety requirements, especially in the vicinity of accommodation areas. Risers should be routed to minimise exposure to marine operations, falling objects and similar hazards.

There is also guidance which requires specific hazard identification and analysis techniques to be applied at the design stage. The DnV Guideline No 1-85 for subsea production systems [3] states that a failure effect analysis is to be carried out. The analysis should deal with the most probable failures, their probabilities and consequences. The result of the analysis should govern the design and content of the operation and in service inspection and testing manuals.

Although hazard and risk studies have been performed by operators for certain pipelines, at present there are no formal requirements for these analyses to be performed for offshore pipelines and subsea systems in the UK. However, this situation is likely to change in the aftermath of Piper A and, as mentioned in the introduction, safety assessments will most probably be required. In relation to pipelines and risers, the areas that will need to be considered include:

- An assessment of the system design and layout
- The performance of the safety critical systems
- Ageing of the equipment and structure
- Recommendations arising from assessment of previous accidents.

It is intended that these safety assessments will be applied as a continuous process throughout the concept, design, construction, major modifications, operation and abandonment stages of an installation's life cycle. For new installations the timing will be targeted to enable the findings of the assessments to be acted upon during the design and construction of an installation. For existing installations a base assessment will be undertaken. Later assessments may then focus on accidents and the lessons learnt, modifications, ageing and abandonment requirements.

STAGES OF THE ASSESSMENT

Assemble System Description

The first stage of the assessment is to assemble information about the pipeline system. It is important not to consider the pipeline as an independent component in isolation from its surroundings. The most common cause of damage to pipelines comes from external sources, see for example References [4-6]. Therefore, knowing the relative position of the pipeline and risers with respect to cranes, platform activities, supply boat anchoring zones, future construction activities, shipping lanes, fishing activities, etc., is as important as knowing the operating temperature or product composition.

Other data that are required for use in risk and reliability analyses are details on uncertainties regarding the definition or description of aspects of the pipeline system including information on variations in geometry, material properties, loading conditions, etc.

Identification of Potential Hazards

The purpose of this stage of the assessment is to assemble a listing of all the potential hazards which might occur [7]. This is basically an intuitive process and so it cannot be guaranteed that all possible hazards have been included. Once a potential hazard has been identified, there is a well developed methodology for assessing it, but this is no help if the hazard has not been listed in the first place. Therefore the identification of areas of vulnerability and of specific hazards is of fundamental importance. There is no single ideal methodology for hazard identification [8,9]. The most appropriate methodology varies to some extent with the type of system and operation under consideration.

Effective hazard identification depends primarily on availability of information on the system and a systematic methodology to ensure the knowledge is applied effectively. There are two types of hazard identification procedure, known as comparative and fundamental methods. Comparative methods rely upon comparing the design with some recognised code or standard such as References [2,10-12]. Checklists are also useful since they can indicate potential hazards, for example, a list of hazards to consider is presented in the discussion document [1]. Another commonly used method to identify potential hazards is to conduct a qualitative review of the system based on experience with other identical or sufficiently similar systems [8]. An important advantage of these methods is that the lessons learnt over the years are available to be used.

Important sources of information on potential failure modes are reports of incidents which have caused damage to pipelines and risers in the past, such as those detailed later in this paper. However, it is important not just to rely on past incidents when compiling a list of potential failure modes. Because of the low level of pipeline and riser operating experience, which for the North Sea is less than 30 years, it is possible that other types of failure, characterised by relatively low probabilities of occurrence, have not yet been observed but could potentially occur. Similar problems occur when novel equipment, components and methods are used.

Such potential failure modes may be able to be identified from a wider study of offshore incident data from other parts of the world, from incidents involving land pipelines [6], or from other industries which use similar equipment. However, care must be taken to check the relevance of data from sources such as these. More formalised methods for identifying hazards are failure mode, effects and criticality analysis (FMECA) [8,13,14] and hazard and operability studies (HAZOP) [7,8]. With these techniques it is possible to identify and evaluate ways in which a system could fail in a systematic way. Fault trees and event trees can also be used for hazard identification.

Analysis of Consequences

While the purpose of the hazard analysis is to identify and quantify the failures which could possibly lead to a critical event, the consequence analysis estimates how potential hazards may affect the system. Separation of the analysis of the hazards from the analysis of the consequences depends on the choice of critical event.

The basis for estimating the expected consequences of a particular hazard often is historical data. However, it is most usual for the consequences of a particular event to be calculated. Analysis methods are often available which can be used in the assessment of consequences. The consequences of various types of defects typical to pipelines, such as material [15] and weld defects [16,17,18], mechanical damage [19,20] and wall thinning [21], can be evaluated using engineering and mechanics of failure theories. Also, certain of the identified failure scenarios, such as upheaval buckling or development of critical spans, may also be assessed by considering the appropriate design data, operational conditions and mechanics principles in the context of limit state and reliability analysis [22].

In some methods the consequence analysis alone is sufficient to enable to comparison of design options. The assumed 'worst case' failures are assumed to occur and the consequences calculated for the different design options. Analyses like this have been performed in the decision of where to locate subsea safety valves. The closer to a platform the valve is the less inventory there is available in a release, because of the shorter length of pipe to the valve and the quicker closure time. However, if the valve is too close then a leak the other side of the valve could impinge upon the platform. In this type of analysis it is important to realise that the quantification of the effects and consequences of incidents also have uncertainties associated with them, even where the physical processes are understood.

Determination of Expected Frequency

In quantitative analysis, the frequency (or probability) of critical events is calculated from statistical data, or where there is insufficient data, from detailed causal analysis. If sufficient data are available from accident statistics or experience on a similar system, the rate can be estimated for each appropriate failure cause [23,24]. Care and judgement must be exercised, however, since accident statistics and descriptions from similar systems may not be appropriate if there are differences in the standards, design, operation, or environmental conditions of the systems being compared.

The estimation of failure probabilities appropriate to the system under study can also be achieved through the use of reliability modelling, for example, through the use of event tree, fault tree and Monte Carlo analyses [7,8,25] Such analysis involves the identification, assessment and modelling of the factors which can lead to a failure. Statistical data can then be generated subjectively, based upon engineering knowledge and experience and available data, when directly relevant historical data are not available. In this way, it is possible to estimate the probability of occurrence of incidents, such as anchor dragging or ship impact with internal risers, etc, or to estimate the probable extent of corrosion, growth of defects, etc.

To be of value a safety assessment should be carried out in accordance with a discipline which ensures that the assumptions made in the assessment are those made in the design. This point is of fundamental importance, since the whole assessment depends on the basic assumptions. Therefore it is important that the assessment is not considered a separate activity and people with the knowledge of the system under study are involved throughout the assessment.

Evaluation of Risk

The final step of the assessment is to combine the results from the hazard and consequence analyses to determine the risks associated with the different hazards. An assessment used to demonstrate the suitability of the chosen configuration at the design stage, and subsequently to predict and monitor performance, pre-supposes that target levels of safety and acceptability have been reached. This is an area of difficulty, as described in the introduction.

It has been argued that the numerical definition of acceptable risk can be achieved by considering the risks that people accept in everyday life. Alternatively available historical data can be used as the basis for setting up target failure probabilities for use in design, assuming that past performance has been acceptable. A target could then be chosen to ensure an order of magnitude improvement over historical rates [22,26].

But these approaches do not necessarily take full account of all uncertainties in the analysis procedure. Such uncertainties limit the extent to which a meaningful quantification can be made of the likelihood of accidents, and more so the probability of harm to people. However, the assessments can be used in a comparative sense to enable decision making, especially in cases where choices have to be made within limited resources. The main benefit of estimating risk lies in the achievement of of detailed understanding of the engineered system and the implications of various siting and technical options [26].

HISTORICAL DATA

It has been described how reports from past incidents can be used at various stages of an assessment. They can be used to identify potential failure modes and their consequences, estimate the probability of occurrence and develop and verify reliability models. However, the application of risk and reliability analysis benefits from the availability of data describing all aspects that govern the performance of the system. Therefore information is required regarding:

- validated historical or test data on component and systems failures;
- rates of occurrence of initiating events;
- data on consequences of hazardous events.

This information required may be readily available in published reports and journals relevant to the particular subject or from existing databases compiled during other related studies. However, the best source of data for the offshore industry comes from the operators and appropriate government departments. Examples of databases relevant to the offshore industry that have been assembled are the OREDA project, phase I of which was published [24] and the ship collision database compiled for the UK Department of Energy [27].

A database of incidents of damage to pipelines and risers in the North Sea is currently being compiled for UKOOA by AME. The main sources of incident data that have been used in this study are:

- o Operator Incident Reports
- o Regulatory Authority Incident Records
- o Operator Commissioned Reports
- o Commercial Databases
- o Published Data from confidential sources
  - Pipeline reliability studies
  - Incident surveys
  - DnV reports and papers
  - Papers and reports concerned with repairs
- o Trade Journals
- o Lloyds List
- o AME Database

When the study is complete, it is intended to release overall numbers of incidents and develop models which can be used to predict failure rates for pipelines. The collation and interpretation of historical data are also discussed in References [28,29]. The next section describes hazards that have caused some degree of damage to pipelines and risers.

HAZARDS TO PIPELINES AND RISERS

Impact Damage from Anchoring Vessels

The largest cause of damage to operating pipelines has been impact from trawl gear or vessel anchors. These account for about half the number of reported cases of damage and cases resulting in leakage.

Pipelines can suffer deep cuts in the concrete due to the sawing action of a mooring line resting across the pipe. A dragging anchor can hook a pipeline and displace it or sever it completely, and no reasonable amount of burial will guarantee protection in this case because anchors can penetrate too deeply into the seabed. Vessels grappling for a lost anchor or dropping anchor too close to a pipeline, particularly in the congested areas close to a platform, can hook the line or hit it. Despite the fact that vessels operating in the vicinity of pipelines are under the control of the platform operator, for example supply boats, the standby boat or construction vessels working on a new line, most incidents of damage result from the activity of one of these vessels. Therefore, historically and statistically a pipeline is most at risk in the safety zone. Anchor damage by a vessel in the open sea has occurred, but is more rare.

Impact Damage from Fishing Gear

A heavy trawl board can scratch or spall concrete and scuff the coating of a flexible and, if the trawl board becomes trapped behind the pipeline, the line can be dragged sideways. Incidents of actual hooking, when the board is permanently trapped, are few but boards have been retained for long enough to displace a pipeline. Very small lines, particularly 2-4" methanol lines piggy backed to a large gas line, have been entirely severed when a trawl board has hooked the line. Small fittings such as bleed valves on a branch connection can be torn off if snagged by a net, resulting in a leak.

Ship Impact

Vessel operations close to platforms involve considerable control and, if this is lost or an error of judgement is made, then a vessel can hit and dent or buckle a riser. Supply boats loading and unloading have been the most common cause of damage to risers [27]. Difficulties of control have been the main causes of vessel impact and not weather conditions.

Considering the possibility of actual contact being so rare, several wrecks have occurred close enough to operating pipelines to cause concern and at least one has landed on and damaged a line.

Dropped Objects

The possibility of dropping equipment off a platform, such as containers, drilling tubulars or scaffolding, onto a pipeline is often investigated. However very few cases of damage have been reported, far fewer than damage from anchors or trawl boards coming into contact with the pipeline.

Material and Welding Defects

Although considerable control is exercised over the manufacture and welding of steel, leaks have occurred due to cracks in both the welds and the main body of pipelines and risers. About a quarter of reported leaks in the North Sea have been due to these defects.

Corrosion

Corrosion has resulted in damage to a number of pipelines and accounts for about a fifth of the incidents which have resulted in leakage in the North Sea. Early risers were coated in bitumen and concrete like the pipelines. However in the splash zone, a far more corrosive environment than the seabed, this was inadequate and cases of external corrosion have been common, although only one has resulted in rupture of a line. Internal corrosion problems can arise when operating conditions or the product carried are changed and the corrosiveness of the new medium is not properly understood. For example, one incidence of internal corrosion damage arose because the operating temperature was increased, accelerating the process.

Design Faults

A pipeline lying on the seabed is subjected to the effects of waves and currents and, in certain cases, pipelines have suffered damage to weight coating. This is most likely when the concrete used for weight coating is low strength and poorly reinforced, as was the case in some of the early pipelines. More recently the structural qualities of weight coating have been improved and, with a greater understanding of hydrodynamic forces acting on pipelines, the problem of instability has become rarer.

Similarly, pipelines have been buried but their relative density has been such that they have risen the sediments and reappeared on the surface.

Pipeline expansion has caused a number of incidents and pipelines have expanded towards the platforms, displacing risers and requiring remedial action such as braces to be removed or the risers to be cold sprung. Recently, upheaval buckling has come to the attention of designers, especially following its occurrence. Upheaval buckling is the natural consequence of a hot, pressurised pipeline being partially restrained by soil resistance but free to expand upwards to some extent. Compressive forces generated by the high temperature cause buckling just as a long strut under axial load will buckle.

Fitting Failures

Fittings such as flanges, connectors and valves are well designed and their location selected carefully, but leaks have still occurred.

Clamp bolts can come loose, due to very bad weather or inadequate design, causing the riser to drop. This has resulted in a pipeline buckling and a leak at a flange.

Pig traps have suffered material defects, resulting in cracking in the body of the fitting, and leaks at seals. However the most serious incidents related to pigging have occurred when the pig trap was at higher than ambient pressure when the door is opened. Pigs have been lost from a trap and some injury sustained because of this.

Construction

Trenching has resulted in damage to pipelines. A number of the earliest pipelines to be installed were damaged by trenchers to the extent that they split during hydrotest. Once this problem was recognised and attributed to poor control of the position of the trencher efforts were made to improve monitoring equipment which would ensure the trencher stopped working if it approached too closely. However, pipelines are still occasionally scraped by trenching equipment.

CONCLUSIONS

This paper has presented the main elements of hazard analysis, part of Formal Safety Assessments [1], as they would be applied to pipeline design. The success of these assessments will depend upon the work not being seen as additional to the engineering effort but as an integral part of it. They need to be performed by engineers with knowledge of the type of system under study or by engineers working in close co-operation with those responsible for the design or operation of the system.

This paper has shown the value of historical data in these assessments. A database of all incidents of damage that has occurred to pipelines and risers in the North Sea is currently being compiled by AME in a project funded by UKOOA. The main sources of incident data that have been used in this study are operator and regulatory authority incident records. Potential hazards and failure modes can be identified and assessed using this historical incident data and reliability modelling techniques.

However, it is also apparent that there is a need for the collation of data concerning pipeline performance and condition, such as is done by NPD [30], and variabilities in design and operation parameters. Using these data it would be possible to examine in more detail the factors which contribute to damage incidents and those which help avert them. The level of sophistication and data availability for the Formal Safety Assessments required in order to achieve improvements in these areas could be identified.

REFERENCES

1. Department of Energy (DEn). Offshore Installations Formal Safety Assessments, Discussion Document, London, 1989.

2. Institute of Petroleum (IP6). Pipeline Safety Code, Model Code of Practice, Part 6, 4th Edition, London, December 1982 and 1986 Supplement.

3. Det Norske Veritas. Safety and Reliability of Subsea Production Systems, Guideline No 1-85, April 1985.

4. E & P Forum. A Review of Submarine Pipeline Performance in the North Sea 1975-1982, Report No. 102, London, October 1984.

5. Cannon, A.G., Lewis, R.C., and Scrivener, C. The Reliability of Pipe Systems Operating in the British Sector of The North Sea; Paper 4A/R, Reliability 85 Conference Proceedings, UKAEA, 1985.

6. Simpson, S. Accidents and Leakage, A Statistical Review, Pipelines and the Offshore Environment Seminar, London, February 1983.

7. Lees, F.P., Loss Prevention in the Process Industries, Volumes 1 and 2, Butterworth 1980.

8. Parry, S.T. A Review of Hazard Identification Techniques and their Application to Major Accident Hazards, SRD R 379, UKAEA 1986.

9. Lowe, D.R.T. and Solomon, C.H. Hazard Identification Procedures - International Study Group on Risk Analysis, I. Chem. E. Symposium Series No. 80, Harrowgate, 1983.

10. Department of Energy (DEn). Submarine Pipelines Guidance Notes, London, October 1984.

11. Det Norske Veritas. Rules for Submarine Pipeline Systems, Oslo, Norway, 1981.

12. Det Norske Veritas. Rules for the Design, Construction and Inspection of Submarine Pipelines and Pipeline Risers, Oslo , Norway, 1976.

13. BS 5760 Reliability of systems, equipments and components Part 3: Guide to reliability practices: examples 1982

14. US Military Standard MIL-STD-1629, Procedures for Performing a Failure Mode, Effects and Criticality Analysis, November 1980.

15. Smith, L.M. and de Waard, C. Selection Criteria for Materials in Oil and Gas Processing Plant, U.K. Corrosion '87, Brighton, 1987.

16. Department of Energy. The Integrity of Offshore Pipeline Girth Welds, Offshore Technology Report, OTH 86 233, HMSO, London, 1986.

17. Maxey, W.A., and Eiber, R.J. Fractures in Pipeline - Main Influencing Factors, Presented at Fracture in Gas Pipelines Seminar, Sponsored by CBMM and CNIICHERMET, Moscow, March 1984.

18. Tenge, P. and Karlsen A. Influence of Weld Defects on Low Cycle-High Strain Fatigue Properties of Welds in Offshore Pipelines, Norwegian Maritime Research, No. 3, 1977.

19. Jones D.G. The Significance of Mechanical Damage in Pipelines, 3R International, July 1982.

20. Maxey, W.A. Analysis of Outside Force Damage to Pipelines, Oil & Gas Journal, May 18, May 25, June 15, 1987.

21. Andersen, T.R. and Sontvedt, T., Monitoring and Prediction of Internal Pipeline Corrosion, Conf. on Pipeline Integrity Monitoring, Aberdeen, 1986.

22. Ellinas, C.P., Walker, A.C., Palmer, A.C., and Howard, C.R. Subsea Pipeline Cost Reductions Achieved Through the Use of Limit State and Reliability Methods, Offshore Petroleum Technology Conference, IBC, 1989.

23. Mann, N. R., et al. Methods for Statistical Analysis of Reliability and Life Data, J. Wiley & Sons, New York, 1974.

24. OREDA - Offshore Reliability Data Handbook, 1st edition; Pennwell Books, 1984.

25. Henley, E.H., and Kumamoto, H. Reliability Engineering , Prentice-Hall Inc., Englewood Cliffs, N. J., 1981.

26. The Royal Society. Risk Assessment, A Study Group Report, London 1983.

27. Department of Energy, Study on Offshore Installations: Protection Against Impact, Offshore Technology Report, OTI 88 535, HMSO, London, 1989.

28. Williams, K.A.J. and Ellinas C.P., Use of Databases in Offshore Engineering, Reliability '89, Brighton, 1989.

29. Williams, K.A.J. Application of Risk Assessment to Pipeline and Subsea Systems, Short Course in Offshore Pipeline Engineering, IBC, London, October 1989.

30. Hokstad, P. Reliability Data for Subsea Pipelines, Report No STF75 A89037, SINTEF, 1989.

Part IV

Deepwater Applications

DEVELOPMENTS IN HYPERBARIC WELDING TECHNOLOGY FOR PIPE LINE REPAIRS BEYOND 600 msw.

J.F. Dos Santos, P. Szelagowski, H. -G. Schafstall and G. Schultheiß

GKSS Forschungszentrum Geesthacht GmbH, Max-Planck-Str., D-2054 Geesthacht, Federal Republic of Germany.

ABSTRACT

The present study discusses the main aspects involved in the selection of welding consumables as well as shielding gas flow and compositions used to produce weldments at 600m and 1100m water depth. The application of these concepts has been illustrated with results from tests carried out at the above mentioned depth range. Also described are the components and mode of operation of a fully automatic system providing an adaptive loop control of welding operations performed under hyperbaric conditions.

1.0. INTRODUCTION

For much of the world, especially in the Gulf of Mexico and Brazil, the best potential for major new oil and gas reserves lies beneath the deep water of the continental slopes, or in other words,

Volume 24: C. P. Ellinas (ed.), Advances in Subsea Pipeline Engineering and Technology, 281–299.

at depths below 450msw. The difficulties involved vary from place to place with environmental conditions, but 450msw to 500msw is a significant depth level since it is about the present limit for ambient pressure diving.

On the other hand, the fall in oil prices has substantially reduced the income of the offshore industry which has been particularly reflected in a decline of the R&D activities. This situation is particular evident in the case of underwater welding technology. The substantial costs and manpower required for the development of mechanized and fully automated welding technology cannot apparently, at present, be afforded by the majority of those companies either requiring or providing such technology.

Hence, in order to adress the problems posed by repair and maintenance at water depths below the limits of manned intervention, GKSS Research Centre Geesthacht GmbH has set up a R&D Programme to develop sensor aided, fully automated robot based underwater welding systems.

The present study has been divided in three main parts. Initially, a brief discussion has been conducted on high deposition welding processes and consumables for hyperbaric welding. In the second part, the most relevant investigations concerning the control of weld metal chemistry of flux cored wires deposited under hyperbaric conditions have been introduced and discussed. The application of these concepts to deep waters is then illustrated with results from mechanical tests, carried out on fully welded joints, produced in the pressure range of 600msw to 1100msw. Finally, a description of the components and mode of operation of a fully automatic system providing an adaptive close loop control of welding operations performed under hyperbaric conditions is presented.

2.0. HIGHER DEPOSITION RATE PROCESSES FOR HYPERBARIC WELDING OPERATIONS

A general overview of the historical development of hyperbaric welding indicates that as the working depth increased and the code requirements became more and more stringent, GTAW was established as the most adequate process for root and eventually hot passes. However, the slow rate of welding and the restricted deposition rate, characterisitcs of the GTAW process, seemed to be a severe handicap, particularly in the case of filler and cap passes. Among the higher deposition rate welding processes which could be considered for hyperbaric use, the solid wire gas metal arc welding (GMAW) and flux cored arc welding (FCAW) seemed to be the obvious choices.

A series of early studies on the properties of consumable welding arcs (solid wire GMAW) operating under hyperbaric conditions showed that the increased thermal conductivity of gases causes constriction of the welding arc and an elevation of the potential drop along the arc column which is approximately proportional to the square root of the ambient pressure [Matsunawa and Nishiguchi, 1979]. The gradual concentration of cathode roots under the arc with increasing pressure leads to substantial vaporization of the base plate. This metal vapour "cloud" is apparently caused to flow away from the surface in the form of vapour jets or plasma streams. This phenomena is considered to be responsible for the accentuated arc instability and increased spatter and fume generation [Burrill and Levin, 1970].

Originally, in the case of GMAW, the restricted performance of conventional welding power sources was considered inadequate to cope with the demands imposed by hyperbaric environments. However, the introduction of microprocessor based electronic power sources has allowed a better definition of process phases which control process behaviour and therefore, stability. Hence, dynamic as well as static process parameters can be varied within high precision levels and reproducibility, at extremely fast response rates. The closer control of metal transfer made possible by such welding power sources are presently being applied to hyperbaric GMAW with varying degrees of success. In any case it seems apparent that GMAW is probably one of the most promising process for hyperbaric use, particularly in the case of mechanised and/or automated applications. Nevertheless, some of the fundamental and practical aspects of metal transfer still require clarification.

The wire/flux concept apparently offered a number of advantages over the bare wire GMAW. Metal/flux combinations are more thermally efficient and fluxing ingredients can be added which improve arc ionization and promote stable metal transfer. An adequate flux formulation and shielding gas selection could offer the possibility of: (a) stable metal transfer due to a balance of fluxing ingredients and gas shield; (b) control of weld metal chemical composition by manipulation of the shielding gas composition and balanced additions of deoxidants; (c) higher heat input levels applied to the workpiece offsetting heat losses due to the effect of pressure; (d) higher metal deposition rate in all welding positions, and; (e) higher tolerance to parametric variations. Additionally, the presence of a slag system, to support and mould the molten metal would improve the bead contour being also of great assistance to positional welding.

All the above mentioned advantages, which are based on fundamental principles of hyperbaric welding technology, encouraged, as early as 1975, the development of "dedicated" flux cored wires which, since then, have been extensively used both offshore and for research purposes.

3.0. FACTORS CONTROLLING WELD METAL CHEMISTRY AND MECHANICAL PROPERTIES IN HYPERBARIC FCAW

According to the literature it could be assumed that ambient pressure affects the weld metal chemical composition through three basic mechanisms: (a) as a result of high ambient pressures gas absorption is increased as well as the break even temperature for Si and Mn control of oxidation reactions; (b) the reduction in cathode and anode spot areas resulting from arc constriction, leads to an increase in energy density and hence higher temperatures on both regions; (c) arc constriction and increased energy densities affect also the resulting bead geometry, particularly the depth of penetration. Since the separation of deoxidation products is related to the molten metal flow pattern, it is possible that different distributions of "hot" and "cold" regions (as proposed by Grong et al., 1986) in the weld pool could lead to variations in inclusion retention levels (when compared to those observed under similar circumstances in atmospheric conditions).

The mutual interactions among the different mechanisms and their overall complexity has not encouraged comprehensive theoretical analysis of such systems. The literature reports however, a series of investigations in which the effects of shielding gas flow and composition on the weld metal metallurgical and mechanical properties has been investigated. In the following items in this section a discussion of these works is presented.

3.1. Experimental Conditions

The flux cored wire used in the investigations hereby reviewed consists on a seamless basic wire and complies to the AWS classification A-5.29/E 80 T5-G. Two main variants of this wire are available: a standard C-Mn type and a C-Mn-1%Ni version with higher strength suitable for fine grained structural steels. Weld metal compositions (0.026% C, 1.09% Mn, 0.98% Ni, 0.009% Mo and 0.026% Al) can be, to some extent, manipulated (through to adjustments in the shielding gas activity) to achieve levels comparable to those obtained at atmospheric pressure and the resulting mechanical properties comparable to those of the parent plate.

The great majority of the investigations discussed in this review have been carried out as a part of the Research and Development Programme of the Institute for Underwater Technology at GKSS Reserach Centre Geesthacht GmbH in the Federal Republic of Germany. The test programmes have been carried out at GUSI (GKSS Underwater Simulation Plant) which is one of the largest and most sophisticated deep diving instalations in the world today. This hyperbaric test facility allows manned diving tests down to a water depth of 600m and unmanned experiments down to 2500m. The R&D Programme in Underwater Welding is divided in six main development fields, which are: (a) Wet Welding; (b) Fundamental Studies on welding processes behaviour under hyperbaric conditions (GMAW, FCAW, GTAW and PLASMA); (c) Manual Welding; (d) Mechanized Underwater Welding (development of orbital welding systems based on GTAW and FCAW processes); (e) Automated Welding Systems (robotic underwater welding), and; (f) Thermal Spraying.

In the present study, results from the Manual Welding and Automated Welding Programmes have been summarized.

3.2. Shielding Gas Flow

Most of the earlier studies on GMAW and FCAW in high pressure environments have used argon as chamber gas. The introduction of nitrogen to the conventional He-O_2 mixtures used as breathing gas in deep diving, made it necessary to use N_2-containing chamber gases to evaluate the effects of eventual contaminations on weld metal properties. Initial investigations carried out under this new environmental gas resulted in weld metal N_2 con-tents in the order of 0.025% to 0.1% at 20bar, due to insufficient shielding action. Such high N_2 levels resulted in severe toughness reduction. The obvious solution would be to increase shielding gas flow rate.

However, although gas flow rate exerts only a small influence on electrical characteristics at atmospheric conditions, this is not the case at high pressures. Under hyperbaric con-ditions, Richardson and Nixon [1985] observed erractic arc and metal transfer behaviour as a result of shielding gas flow instabilities promoted by excessive flow rates. Therefore, a crictical shielding gas flow could be defined based on two criteria: arc stability and shielding efficiency.

However, in practical terms, the "effective shielding action" criterion has to be primarily fulfilled in order to avoid excessive weld metal nitrogen and hydrogen contamination (even if the required shielding gas flow is superior to that needed to satisfy the stability criteria). To this effect, Huismann et al. [1985] have shown that up to 17 bar, the shielding gas flow required to keep contamination below acceptable levels (approximately 0.01% N_2), is nearly a linear function of pressure. It is interesting to observe that in this work, the authors also measured the contamination profile (percentage of chamber gas present in the shield) of different gas nozzels using a capillary mounted in a fixed position beneath the torch. The reported results showed that initially, the contamination level decreased with increasing shielding gas flow rates. However, above a specific flow rate (which varied according to the ambient pressure) the contamination level increased

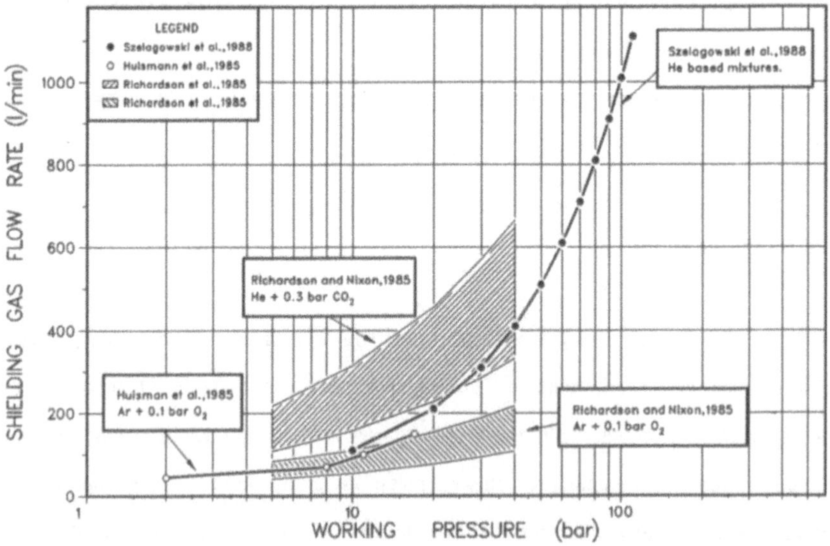

Fig.1 - Suggested shielding gas flow rates as a function of ambient pressure.

drastically. It seems reasonable to assume that this increase in the contamination level is associated with increased turbulence in the shield gas flow and could be related to the arc instabilities reported in the literature.

Szelagowski et al. [1988] reported the results of manual and robotic dry hyperbaric FCAW down to 450msw in which the influence of welding parameters, technique and consumables on overall weldment quality have been investigated. It has been shown that using He-based shielding gas mixtures and different metal transfer modes, a shielding gas mass flow rate equivalent to 10 l/min. at the working depth in question, resulted in weld metal nitrogen contents between 0.008% and 0.015%.

Figure 1 presents a summary of the suggested shielding gas flow rates found in the literature.

3.3. Shielding Gas Activity and Weld Metal Oxygen Content

Under effective shielding action, oxygen absorption from the arc atmosphere depends largely on plasma jet phenomena; flow rate, composition, dissociation and ionization properties of the shielding gas; and absorption mechanisms at the molten metal-gas interface. It has been conclusively shown that for a given active gas content in the shielding gas, increasing pressure leads to higher oxygen contents in the weld metal [Müller, 1988].

Figure 2 presents the published relationships between oxygen and CO_2 partial pressures (in the shield) and the resulting weld metal oxygen contents, for C-Mn and C-Mn-1%Ni wires, welded with positive polarity [Müller, 1988; Dos Santos et al., 1988; Szelagowski et al. 1988]. As indicated in Figure 2, with the exception of the selected polarity which is common to all specimens, these results cover a wide range of experimental conditions. Despite these differences, only a few specimens presented weld metal oxygen contents between 0.04% and 0.05%. Considering that according to the literature, the optimum oxygen content from a toughness point of view lays between 0.03% and 0.04%, there was reason to believe that such conditions could not be possibly achieved under hyperbaric conditions.

However, a series of ICP-OES analysis of flux samples revealed Ca contents in the order of 45%. In standard gas shielded flux cored wires for atmospheric use, Ca is usually present in the form of CaF_2, CaO or $CaCO_3$. These compounds provide additional shielding gas (fluoride shielding, CO and CO_2), functioning also as slag formers, fluxing agents and arc stabilisers, being therefore also essential for hyperbaric use. As a result of that, a certain amount of oxygen will be always present in the arc atmosphere, originated either from the decomposition of CaO and $CaCO_3$ or through slag/metal reactions involving non-decomposed carbonates or oxides. As a matter of fact, specimens from test plates welded at atmospheric conditions, using shielding gas mixtures with very low active gas participation (Ar-0.1%O_2, Müller, 1988 and He-1.0%CO_2, Dos Santos et al., 1989a), resulted in weld metal oxygen contents around 0.023%, which could then be considered as a "residual" value for this type of wire.

At this stage, it is of particular importance to observe that early experiments, conducted at GUSI using NEGATIVE POLARITY and low $C0_2$ partial pressure in the shield, have produced weld metal oxygen levels below 0.045%. Similar results have been later reported by Müller [1988] using Ar-0.3%O_2 in the pressure range from 15 to 100bar (O_2 partial pressures between 0.05bar and 0.3bar).

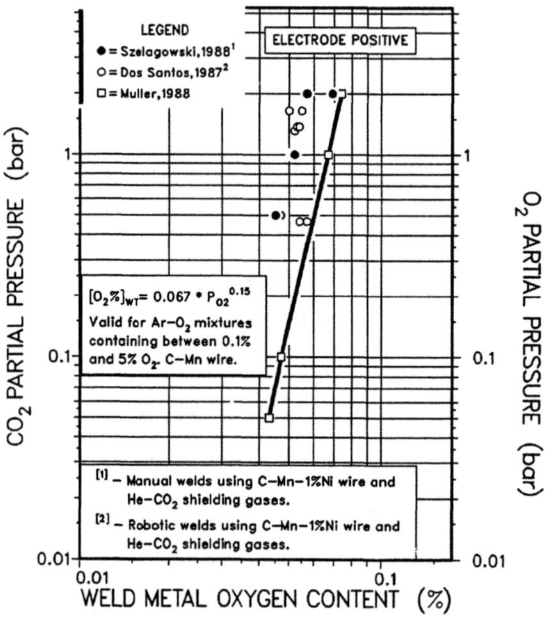

Fig.2 - Relationship between shielding gas activity and weld metal oxygen content.

These results indicated that weld metal oxygen levels below 0.045% could be achieved primarily by the selection of low O_2 partial pressures in the shielding gas and negative polarity. So far, to the best of the authors' kwnoledge, there are no experimental evidence to explain the low oxygen contents obtained with negative polarity under hyperbaric conditions. However, the findings of early works on the physics of consumable arcs under hyperbaric conditions could provide some insight on this phenomena. Matsunawa and Nishiguchi [1979] observed that arc constriction resulted in an increased generation of metallic vapour from the electrode, as a consequence of superheating. On the other hand, fundamental studies on gas/metal reactions [Corderoy, 1980] have observed that longer droplet retention times resulted in lower oxygen absorption. It has been suggested that vaporised manganese and iron would react with atomic oxygen in the vicinity of the droplet. It is therefore probable that the excessive metal vapour generated at wire tip reduces the oxygen supply to the droplet surface resulting in correspondingly lower weld metal oxygen contents.

The results presented in the literature show that low active gas partial pressure in the shield leads to the oxygen levels required for a satisfactory weld metal toughness behaviour [Müller, 1988]. The use of negative polarity has been also considered as a favourable element to achieve these levels. Another factor to be taken into account is the selected active gas. Figure 2 shows that, as expected, the specimens welded under He-CO_2 mixtures presented always lower weld metal O_2 contents.

3.4. Deoxidants and Remaining Alloying Elements

Coe and Moreton [1979] conducted a study where the theoretical relationships of oxygen reactions which take place in the arc and weld pool, their correlations with plate, wire and shielding gas composition, and the resultant weld metal composition were studied. The resulting expressions have then been applied to experimental results as well as to data provided in the literature. In order to further substantiate these results, Moreton and Boothby [1981] conducted additional experiments under hyperbaric conditions, to study the effect of varying partial pressures of the gaseous components. These experiments indicated that the proposed equations could also be applied to high pressure systems requiring however a proper selection of the values for ambient pressure and the dissociation degree of CO_2.

Dos Santos et al. [1989a] applied these equations to selected results obtained in the manned hyperbaric weldig programme at GUSI. In order to adjust the experimetal conditions of this programme (particularly ambient pressure and carrying gas) to the boundary conditions of these equations, some considerations had to be made, especially regarding the reaction temperature. A simple "best fit" procedure using the proposed equations and the available data indicated that a value of 1800°C could be assumed as an approximate value, representative of the experimental conditions reported. The resulting Equation (1) is expressed by:

$$\log\left[\frac{P_{CO_2}}{[\%\,O]^2\,[\%\,C]}\right] = \frac{9175}{T} - 2.304 \qquad (1)$$

where P_{CO2} is the CO_2 partial pressure, [%O] and [%C] are the weight % concentrations of oxygen and carbon in the weld metal and T = 2073 °K (1800° C) respectively. The relation between P_{CO2} and the volume % of CO_2 in the shielding gas (V) can be expressed by Equation 2:

$$P_{CO_2} = \frac{\dfrac{2P(100-\alpha)}{100}}{2+\left(\dfrac{\alpha}{100}\right)+2\left[\dfrac{(100-V)}{V}\right]} \qquad (2)$$

The values of "α" (degree of dissociation of CO_2) for each working pressure are derived using equations which take into account the equilibrium constant of the reaction, the reaction temperature and the ambient pressure.

It was observed that Equation (1) described, to some extent, the relationships between weld metal carbon and oxygen contents, according to pressure and shielding gas activity (Fig. 3). It was observed that, as suggested by Moreton and Boothby [1981], the carbon content of hyperbaric weld metals is undoubtedly higher than those reported for specimens welded under similar conditions at atmospheric pressure. Moreover, the reported results also indicate a limited but dicernable tendency of higher carbon contents for increasing effective partial pressures of CO_2 in the shielding gas, accompanied by an equally dicerniable decrease in manganese and silicon contents (Figs. 4 and 5).

In conclusion, Dos Santos et al. [1989a] suggested that in the case of hyperbaric FCAW, for a given working pressure, the increased availability of CO and CO_2 leads to higher carbon

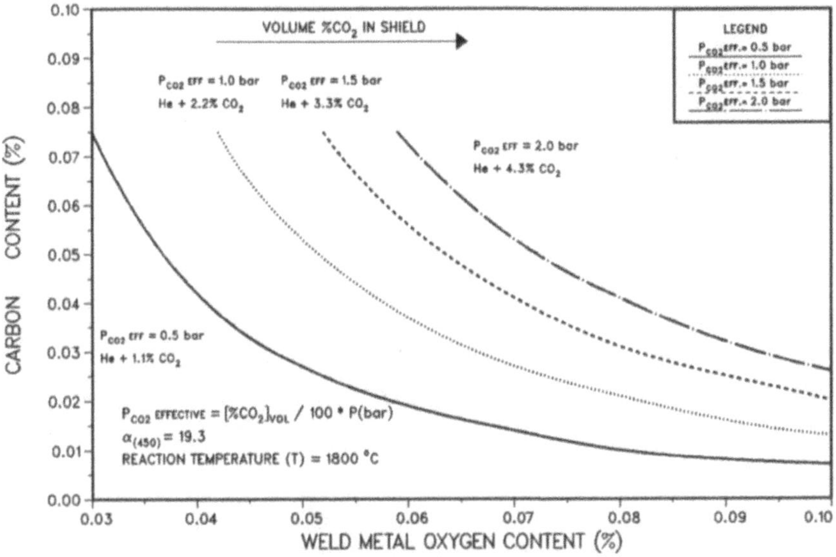

Fig.3 - Effect of shielding gas activity on C and O_2 contents.

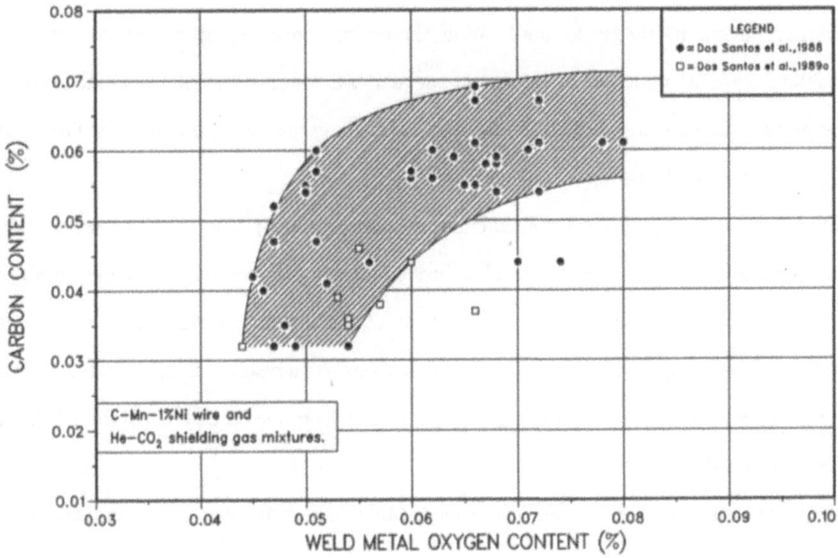

Fig.4 - Relationship between weld metal O_2 and carbon content.

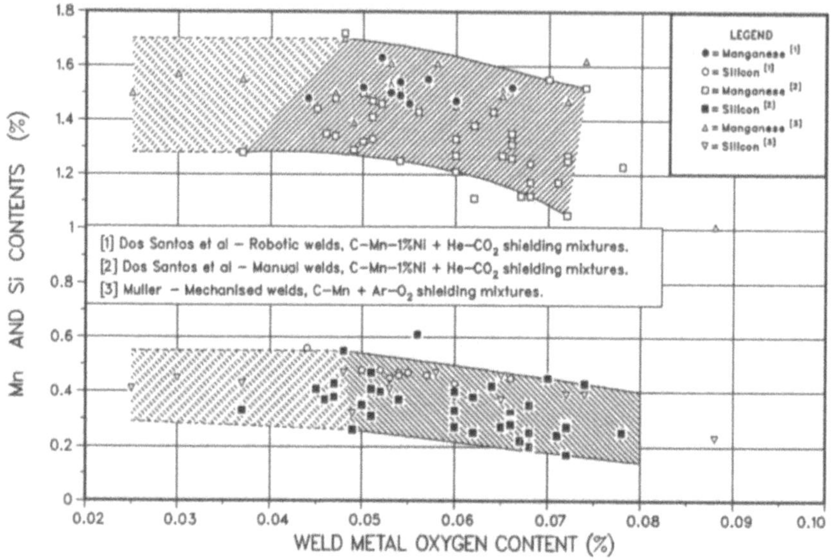

Fig. 5- Relationship between weld metal O_2, Mn and Si levels.

equilibrium levels in the weld pool. With increasing pressure, upon reaching a critical temperature, carbon oxidation in the liquid metal is no longer thermodynamically favoured, giving place to deoxidation through Si and Mn, confirming the model proposed by Grong et al. [1985] for hyperbaric SMAW.

As far as the remaining alloying elements are concerned, the literature has not yet provided a systematic analysis of the reported results. In some cases the recorded variations in alloying contents are not significant or else, the amount of specimens produced under comparable conditions is not enough for a reliable statistical analysis. As a result of that, variations in Nb, V, Ti and Al have been usually attributed to dilution effects from the base plate [Müller, 1988; Dos Santos et al., 1988].

In all the investigations carried out using the C-Mn-1%Ni wires, there is no indication of Ni losses, regardless of pressure, shielding gas or metal transfer regime.

3.5. Properties of Weldments Obtained in the Depth Range Between 600 and 1100msw

Fully welded test plates (20mm thick, StE445.7 TM steel) were produced as part of the 3rd Series of Tests of the Robotic Underwater Welding Programme. Welding parameters with positional capabilities (i.e., low heat input) were adopted. He-CO_2 mixtures providing CO_2 partial pressures of 0.5 bar abs and 1.0 bar abs. were employed. Shielding gas flow strategy was based on that proposed by Szelagowski et al. [1988].

Weld metal N_2 contents (which function as an indicator of contamination levels) were observed to be restricted to the expected range of 100ppm $\pm$ 10ppm. Preliminary evaluation of welding voltage and current traces revealed that the shielding gas flow settings used did not significantly influenced process stability. Oxygen contents varied between 0.053% and 0.065% for shielding gases with a CO_2 partial pressure of 0.5 bar abs. to 1.0 bar abs. regardless of the working pressure. Charpy impact energy at -20° C obtained at the studied working pressures (60 to 110 bar abs.) varied from 52J to 107J and therefore, comply with the standards conventionally employed in offshore constructions. Figure 6 summarises some of the obtained results.

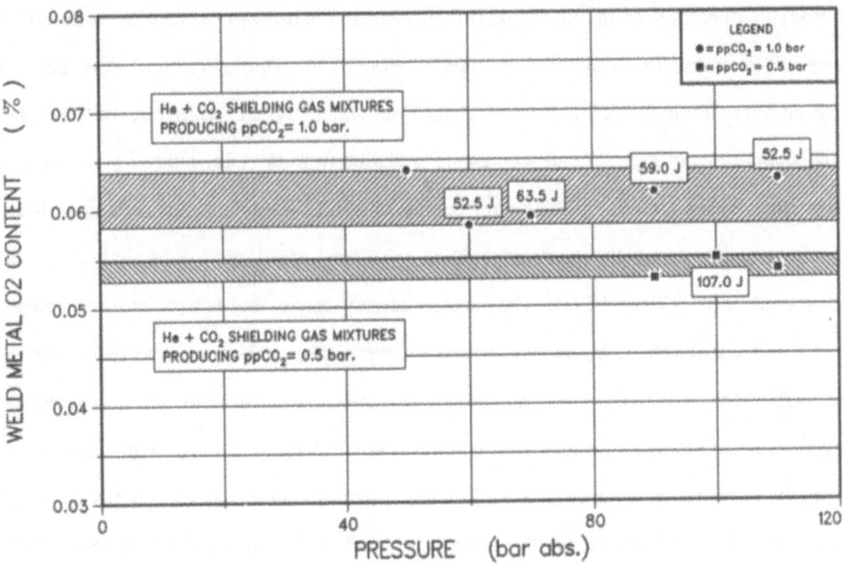

Fig. 6 - Weld metal oxygen content as a function of the shielding gas activity for the specimens welded in the pressure range from 50bar to 110bar. Figures inside the squares indicate the Charpy impact energy at -20° C.

4.0. AUTOMATED PROCESS CONTROL FOR MECHANIZED UNDERWATER WELDING

In order to implement adaptive loop control in hyperbaric welding, a system composed of a sensor controlled second generation robot with six degrees of freedom, an electronic analog welding power source, a seam tracking unit and a central management unit (system processing unit) has been developed. The system processing unit, based on a 68000 MC processor including a VMEbus system, controls the whole data transfer among peripheral components. The data management is performed by a 10Mhz, 16 bit Virtual Memory Microprocessor (CPU1) with a SCSI and VMEbus interface, and a 20MHz 32 bit Virtual Memory Microprocessor (CPU2) with a 2MB dual ported RAM and a floating point co-processor. Additional elements of the system processing unit are: a 32 channel data aquisition board including a 12 bit A/D interface with direct access to the VMEbus and optical isolation between the VMEbus and the I/O section, a 70MB Winchester disc drive and a 655kB floppy disc drive. A full description of the robot unit and seam tracking system used can be found in the literature (Dos Santos et al., 1989b).

The mode of operation of the above described process control system can be summarized as follows: the seam tracking camera is mounted in front of the welding torch. As the robot's end effector moves along the joint groove, images generated by the seam tracking camera are processed and the root gap value as well as the deviation from the joint centre line are conveyed to the system management unit. The data generated by the image analysis unit as well as the values for welding parameters (current, voltage, wire feed speed, etc.) are sampled at a 1MHz frequency and read by the CPU1. The same processor carries out a comparison between the expected and actual values. In case differences between these values are detected a new set of correct values for torch dislocation and welding parameters are generated. These parameters are changed according to synergic type relationships and are determined by a mathematical model existing in the working memory of CPU1. These equations make use of a series of constants stored in the form of matrix in the memory of CPU2. The interaction between CPU's is performed by the VMEbus and the dual ported RAM. The new calculated parameters are transmitted by a current loop interface to the welding power source. The actual weaving parameters are transmitted to the sensor input of the robot control unit through a D/A interface.

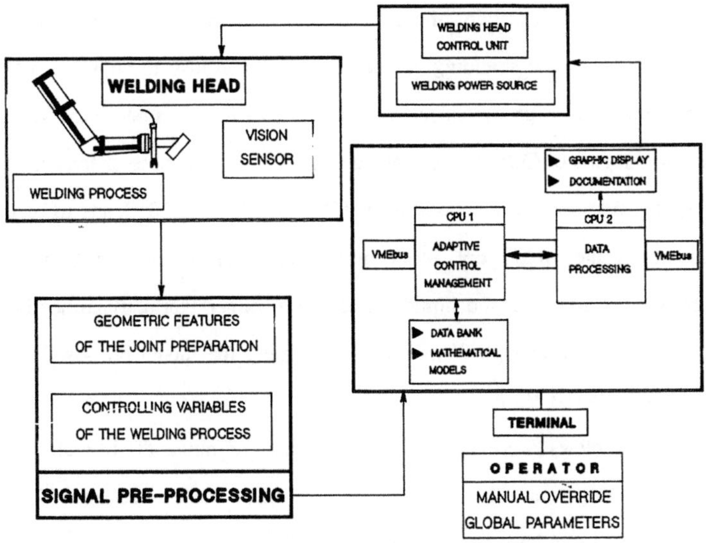

Fig. 7 - Adaptive control for mechanised underwater welding.

5.0 CONCLUSIONS

A literature review on the factors affecting the weld metal chemistry of hyperbaric flux cored arc welds has been carried out. The following general conclusions could be drawn:

(1) The concept of optimum shielding gas flow conditions for hyperbaric FCAW should be based on two criteria: arc stability and effective shielding action. The latter should primarily be fulfilled to avoid excessive weld metal contamination by hydro-gen or nitrogen (in the case of He-O_2-N_2 chamber mixtures). Due to its physical properties, Ar based mixtures require comparatively lower flow rates for a equivalent "protection" levels.

(2) It has been conclusively shown that for a given shielding gas composition, increasing ambient pressures lead to higher weld metal oxygen contents. Low active gas partial pressure in the shield (in the order of 0.1bar to 0.3bar) leads to oxygen levels required for a satisfactory weld metal toughness behaviour. The use of negative polarity and CO_2 containing mixtures have been also shown to favour low oxygen (between 0.03% and 0.045%) levels.

(3) Equilibrium expressions describing the dependence of C and oxygen levels on the partial pressure of CO_2 and O_2 in the shielding gas have been shown also to apply to high pressure systems requiring however, a proper selection of the values for pressure, reaction temperature and the dissociation degree of CO_2. It has been suggested that in the case of hyperbaric FCAW, for a given working pressure, the increased availability of CO and CO_2 leads to higher carbon equilibrium levels in the weld pool. With increasing pressure, upon reaching a critical temperature, carbon oxidation in the liquid metal is no longer thermodynamically favoured, giving place to deoxidation through Si and Mn, confirming the model proposed by Grong et al.[1985] for hyperbaric SMAW.

(4) Fully welded test plates were obtained in the pressure range 60 bar abs. to 110 bar abs..Charpy impact energies at -20° C of these specimens varied between 52J and 107J, complying therefore with standards conventionally used for offshore constructions.

REFERENCES

Burrill, E.L. and Levin, M.L.: "The Effect of Pressure on the MIG Welding Arc", Proceedings of an International Conference on Gas Discharges, London, November 1970, 595-599.

Coe, F.R. and Moreton, J.: "The Chemistry of Oxygen in Gas-Shielded Welding of Steel", The Welding Institute Research Report No.: 95/1979, Cambridge, 1979.

Corderoy, D.J.H., Wills, B. and Wallwork, G.R.: "Gas/Weld Metal Reactions in MIG Arc Plasma", Proceedings of an International Conference on Weld Pool Chemistry and Metallurgy, London, Paper 12, 15-17 April, 1980, 147-153.

Dos Santos, J.F., Szelagowski, P., Schafstall, H.-G. and Hensel, D.: "Mechanical and Metallurgical Properties of Robotic Underwater Welds Performed Within a Depth Range of 100 to 1100 msw", Proceedings of the 7th OMAE Conference, Houston, Volume III, 7-12 February, 1988.

Dos Santos, J.F., Szelagowski, P., Küster, K. and Schafstall, H.-G.: "Aspects of Oxidation - Deoxidation Reactions of Weld Metals Deposited Under Hyperbaric Conditions", Proceedings of the 8th OMAE, Den Haag, 19-23 March, 1989a.

Dos Santos, J.F., Kober, K.-B., P.Szelagowski, G. Seeger and G. Schultheiß.: "Results of the Application of a Seam Tracking System in Robotic Hyperbaric Root Pass Welding", Proceedings of a Conference on Welding Under Extreme Conditions, Helsinki, 4-5 September, 1989. Pergamon Press.

Grong, O., Olson, D.L. and Christensen, N.: "Carbon Oxidation in Hyperbaric MMA Welding", Metal Construction, December 1985, 810R-814R.

Grong, O., Siewert, T.A., Martins, G.P. and Olson, D.L.: "A Model for Silicon-Manganese Deoxidation of Steel Weld Metals", Metallurgical Transactions A, 17(10), October 1986, 1797-1807.

Huismann, G., Hoffmeister, H. and Schafstall, H.-G.: "Effect of the Ambient Pressure and Shielding Gas Flow on Hydrogen and Nitrogen Contamination of Hyperbaric TIG and GMA Welds", Proceedings of an International Conference on Welding in Challenging Environments, Ontario, 14-18 October 1985.

Màtsunawa, A. and Nishiguchi, K.: "Arc Characteristics in High Pressure Argon Environments", Proceedings of an International Conference on Arc Physics and Weld Pool Behaviour, Lonodon, 8-10 May, 1979, Vol 1, Paper 8, 123-133.

Moreton, J and Boothby, P.J.: "The Chemistry of Oxygen in Gas Shielded Welding of Steels: Part II", The Welding Institute Research Report No.: 143/1981, Cambridge, 1981.

Müller, L.: "Gefüge und Zähigkeit von ferritischem Schweißgut aus hyperbaren Einsatz", Universitäat Hamburg-Harburg, Dissertation, Hamburg, 1988.

Richardson, I. and Nixon, J.H.: "Open Arc Pulsed Current GMAW - Application to Hyperbaric Welding Operations", Proceedings of an International Conference on Welding in Challenging Environments, Ontario, 14-18 October 1985, 153-163.

Szelagowski, P., Dos Santos, J.F., Loebel, P., and Schafstall, H.-G.: "Considerations on Welding Pro-cedures for Hyperbaric FCAW Down to 450 msw", Proceedings of the 20th Offshore Technology Conference, Houston, Paper OTC 5818, 2-5 May, 1988, 117-125.

GLOSSARY

The definitions listed below have been obtained from the following sources: Welding Handbook, Volume 1, Welding Technology, 8th Edition, American Welding Society; Metallurgy of Welding, by J.F. Lancaster, 4th Edition, Allen & Unwin, London; and The Physics of Welding, edited by J.F. Lancaster for the International Institute of Welding, 2nd Edition, Pergamon Press, Oxford.

anode spot. The portion of the positive electrode within which the electrons are absorbed.

cathode spot. That part of the negative electrode from which the electrons are emitted.

deposited metal. Filler metal that has been added during welding.

deposition rate. Is the weight of weld metal deposited per unit of time by a particular process/consumable combination.

droplet retention time. Time period in which a molten droplet grows at the electrode tip before being transferred to the welding pool.

filler and cap passes. The remaining welding beads deposited after the root pass.

filler metal (filler wire). The metal to be added in making a welded, brazed or soldered joint.

flat position. The welding position used to weld from the upper side of the joint; the face of the weld is approximately horizontal.

flux. Material used to prevent, dissolve or facilitate removal of oxides and other undesirable surface substances.

flux cored arc welding (FCAW). An arc welding process that produces coalescence of metals by heating them with an arc between a continous filler metal electrode and the work. Shielding is provided by a flux contained within a tubular electrode. Additional shielding may or may not be obtained from an externally supplied gas or gas mixture.

flux cored wire. A composite filler metal electrode consisting of a metal tube or other hollow configuration containing ingredients to provide such functions as shielding atmosphere, deoxidation, arc stabilization and slag formation. Minor amounts of alloying materials may be included in the core. External shielding may or may not be used.

gas metal arc welding (GMAW). An arc welding process that produces coalescence of metals by heating them with an arc between a continous filler metal electrode and the workpieces. Shielding is obtained entirely from an externally supplied gas. In the past refered to as MIG welding process.

gas nozzle. A device which directs shielding media (gas).

gas tungsten arc welding (GTAW). An arc welding process that produces coalescence of metals by heating them with an arc between a tungsten electrode (nonconsumable) and the workpieces. Shielding is obtained from a gas. Filler metal may or may not be used. In the past refered to as TIG welding process.

heat input. If "V" represents ar voltage, "I" arc current and "n" is the proportion of arc energy that is transferred to the workpiece, then the heat input rate "q" per unit length of weld is:

$$q = \frac{n\mathrm{VI}}{v}$$

where "v" is the welding speed. This parameter governs heating rates, cooling rates and the weld pool size.

joint (welded joint). The junction of members or the edges of members which are to be joined or have been joined.

metal transfer. Is the manner in which liquid metal flows from the elctrode tip into the weld pool

negative polarity (direct current electrode negative - DCEN). The arrangement of direct current arc welding leads in which the workpiece is the positive pole and the electrode is the negative pole of the welding arc.

OES-ICP. Opitical Emission Spectrometric Analysis by Inductive Coupled Plasma

positive polarity (direct current electrode positive - DCEP). The arrangement of direct current arc welding leads in which the workpiece is the negative pole and the electrode is the positive pole of the welding arc.

root pass. The welding bead deposited on that portion of a joint to be welded where the members approach closest to each other.

shielded metal arc welding (SMAW). An arc welding process that produces coalescence of metals by heating them with an arc between a covered metal electrode and the workpieces. Shielding is obtained from decomposition of the electrode covering. Filler metal is obtained from the elctrode.

solid wire GMAW. A nonstandard term for gas metal arc welding. Used to emphasize the difference between flux cored wire and solid wire continous welding processes.

weaving parameters. The elements defining the geometric characteristics of the transverse oscilalation movement imposed on a welding torch to improve the efficency of deposition or to facilitate root and positional welding.

weld metal. That portion of a weld that has been melted during welding.

welding torch. A device used in semiautomatic, machine and automatic arc welding to transfer current, guide the consumable electrode, and direct the shielding gas.

STEFANO BIANCHI
TECHNICAL DEPT. - MANAGER, ENGINEERING AND STRUCTURAL ANALYSIS
SAIPEM S.p.A. - SAN DONATO MILANESE (ITALY)

"LAYING DEEPWATER PIPELINES IN INTERNAL WATERS (Lake of Como) TECHNOLOGY, TECHNIQUES, ENVIRONMENTAL FACTORS"

INTRODUCTION

In 1988 ACEL, the municipal gas company of the town of Lecco, commissioned a turnkey project involving the installation of three gas lines respectively 2.2, 3.9 and 4.7 Km in length, connecting the towns of Olcio, Dervio, Bellano and Varenna on the left shore of the lake.
The customer needed an off-shore, instead of the classical on-shore, linkup, owing to the particular nature of the lake itself and its surroundings.
Lake Como is, in fact, a typical glacial lake with great, craggy mountains forming its shoreline. Consequently, the solution of laying the lines in rock was discarded by the customer for environmental and economic reasons. The particular morphology, environmental constraints, and inordinate number of caverns and recesses, as well as the difficulties involved in procuring suitable equipment in the area, however, largely influenced the techniques adopted to lay the lines.

Lake Como is a very well-known Italian tourist center whose shores are dotted by numerous towns and villages connected to each other by a large fleet of hovercrafts and ferries.
Moreover, its shores are generally extremely craggy and rocky and drop precipitously below the water level to depths which, for our pipelines, exceed 300 meters.

Volume 24: C. P. Ellinas (ed.), Advances in Subsea Pipeline Engineering and Technology, 301–320.

One of the characteristics of the possible landing sites is that they are for the most part formed by alluvional cones of the torrents issuing from the many valleys high above the lake. These cones are made up of masses of unstable gravel and depend for their existence on the flooding of the torrents.

Therefore, the selection of the landing sites and the routes to reach them turned out to be one of the most important aspects of the project.

All the operations that we shall now undertake to describe took place over the period from November through December 1988.

Planning

The draw backs of the system within which the installation had to be planned may be summarized as follows:

- exceptionally great depths;
- difficulties in obtaining equipment with power ratings of not more than 400 HP;
- mooring problems resulting from steepness of the rocky shoreline;
- the danger of landslides in the areas where the alluvial cones were found;
- the discovery of certain obstructive aspects in the ecological environment that made it absolutely imperative to recover all the material used in carrying out the project.

The classic solution of using barge to lay the pipe, equipped with prefabricated pontoons put together on the spot was, therefore, rejected, owing to the difficulties that would have to be faced in guaranteeing suitable anchorage, in controlling the handling of the anchors themselves without adequate means, in not obstructing the navigation on the lake, and in operating the enormous stinger and/or tensioners required to guarantee the correct of stress in the line.

The Surface-Tow Solution

The problem was solved by prefabricating the strings on land, assembling

them in the water to their planned length, and then moving them to their final destination using the surface-fow method.
Once there, the strings were moored and, after activating the tensioning system, were put in place using a method, patented by Saipem, of mechanically releasing the blocking clamps installed on the floaters.

Once this choice was made, there were several problems that had to be solved, namely:

- the sizing of the line;
- the selection of the prefabrication site;
- the determination of the size of the mooring-points for the assembly of the prefabricated strings;
- navigation and critical timing problems;
- the mooring and protection of the line;
- the setting-up of the line-tensioning system;
- the release of the floaters and their recovery;
- land excavation problems;
- the surveillance and use of the ROV to release anomalous floaters and of the first line tensioning system;
- final testing.

The Sizing of the Line

The size of the line to be such that it would:
- meet the required flow rate;
- withstand the required operating pressure and meet current regulations;
- withstand the amount of stress allowed for during its installation;
- withstand allowed operational stress.

Of these requirements, only the last two were found to be of significance in sizing the line.
For the installation, operational and testing stages, we chose the following criteria:

85% SMYS as the maximum allowable stress rating during installation;
75% SMYS during the operational stage;
90% SMYS during testing.

Placing the lines proved to be an extremely delicate operation, because of the great depth involved and the fact that, since they had to take on a semicircular shape, it was impossible to guarantee a pull on the conduits without creating big complications throughout the entire system.

The use of X65 as the piping material followed upon the discovery that it guarantees a sag-bend stress level, when stretched between two supports, of less than 70% even without exercising any pull (see fig. 9), but, considering only the pulling effect that would arise from the floats, whose axes, at the deepest point, would be at an angle of about 15° to that of the pipe.

On the other hand, mathematical simulations also demonstrated that X65 would be necessary to guarantee the operational stress of the project.

This stress would have become critical, if the testing had been carried out by introducing water into the line, and this is the reason why the tests were made using gas.

Although the amount of free spans that could be predicted by calculations regularly turned out to be correct, this factor, which is typically of the utmost importance where seas and oceans are involved, had no effect whatsoever on the planning of the line, since the current at the bottom of the lake is practically nil.

The analysis outlined above provided the following line specifications:

- outside diameter 6.625 in.
- gauge 7.11 mm
- material X65
- naked piping covered with rock protective netting.

Description of the Calculation Programs Used to analyse the Line

From the preceding considerations, the importance of the calculation programs used in the experimental phase in determining the feasibility of the project, as well as during its execution, becomes clear.

The simulation phase was primarily concerned with all the aspects involved in getting the pipe laid, from the characteristics of the anchoring blocks, to the pinpointing, at the assembly site, of the excavations and final positioning of the completed string.

In all events, the most interesting aspects of this phase of the operation concerned the analysis of:

- the pipe during installation and the release of the buoys;
- the pipe once placed on the bottom;
- the behaviour of the tensioning system during the night station keeping and while being installed;
- the mooring of the line at the time it was being put together at the assembly site.

The first two stages were analysed together, since the release of the pipe from the floats and its positioning on the lake floor occurred at the same time.

As for hydrotesting, on the other hand, which was subsequently discarded, and testing with gas, the pipe was analysed only after being positioned on the lake floor.

The description of the Program

The program, therefore, had to take into account both the introduction of the floats and their behaviour, and the problem of the adaptation of the line to the lake bottom.

How pipe reacts to the effects of floats, which, in our case, were cylindrical, is, in principle, not much different from its behaviour when lying on the bottom.

In fact, the action of a cylindrical float is equivalent to the action of a linear spring whose elastic constant equals gamma times the area, where gamma is the specific gravity of water and the area is the sectional area of the cylinder.

To complicate things, however, two strong non-linearities come into play, namely, that of a float with no thrust and that of a float completely immersed in water.

The behaviour of a pipe resting on the ground can also be simulated by using a non-linear spring, whose non-linearity is essentially attributable to the one-sideness of the contact and the cosinusoidal variation of the section in contact with the bottom, caused by the circular shape of the pipe.

The program can be used to analyse various types of soils and, as a consequence, a series of springs used to simulate them.

The first problem that presents itself in this type of analysis is what is known as earth data preprocessing. In fact, for Lake Como we received a magnetic support containing earth data recorded at a sampling frequency of one every two meters. From this the following operational sequence was built up:

- the creation of a file containing the raw data;
- this file is put through a filtering and smoothing routine to remove possible spikes;
- an automatic check is made to find any holes in the data sequence;
- a routine spline is applied in order to further concentrate the data to a density of one per meter by interpolating the existing data;
  (once this particular job is done, it is possible to analyse the entire profile with each run - see figs. 1 and 2)
- the physical and geometric parameters of the pipe are prepared for each section.

Actually, earth data processing may be performed in two ways:

- via the analysis of hard ground, or
- via the analysis of soft ground.

The two types of analysis are conceptually different only in that hard ground, though infinitely rigid, spring, is just soft earth under particular circumstances. In practice, however, the solution involving hard ground is much more rapid, since it is possible to exploit the linearity of the system of equilibrium equations that define it, once the non-linearity of the springs representing the softness of the ground has been eliminated.

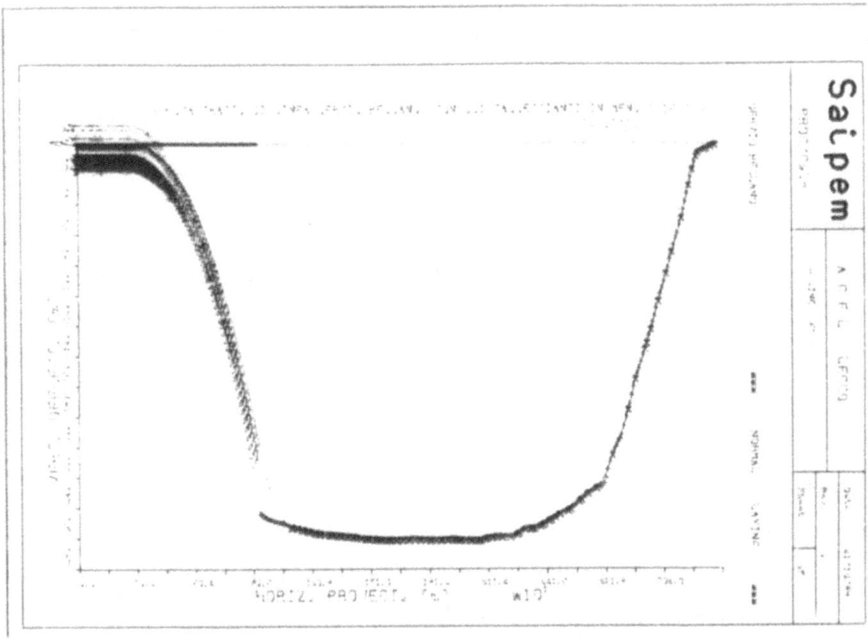

Fig. 1 - Computer Output - Pipe Geometry

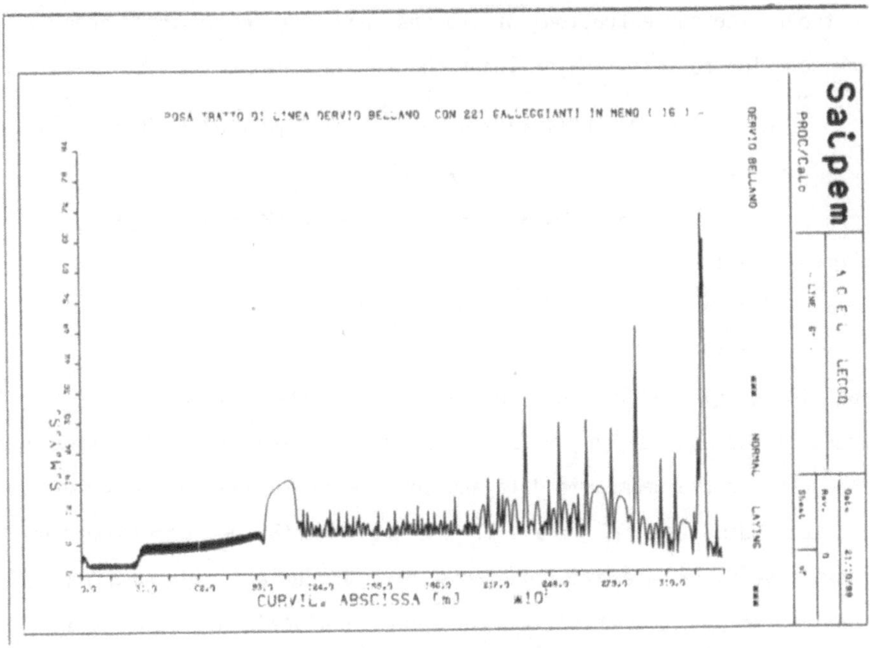

Fig. 2 - Computer Output - Pipe Stress

With Lake Como, the hard-ground version was used in view of the prevalently rocky condition of the area.

In any case, it should be noted that the hard-ground solution is more conservative from the standpoint of stress and free spans.

The Mathematical Model

The mathematical model is based on the integral of the differential equation:

$$EJY'''' \quad -NY'' = -wt$$

where:

EJ = Bending stiffness;

N = Pull;

wt = weight in the water per meter of the pipe (transversal

Y = Pipe deformation

The pipe is divided up into a group of nodes and elements. In each node the equilibrium equations are defined and the springs simulating the ground or pontoons, the concentration of forces and the restraints are set up. The solution to this problem leads to:

$$m1 \times c1 + m2 \times c2 + m3 \times c3 - y1 \times d1 + y2 \times d2 + y3 \times d3 = b1$$

and

$$m1 \times a1 + m2 \times a2 + m3 \times a3 = K2\ (y2 - y20) + b2$$

where: 1 = i -1

2 = i

3 = i +1

and "i" is any node in the schematic representation of the pipe

where c1, c2, c3, d1, d2, d3, a1, a2, a3, b1, and b2 are the coefficients depending on the geometric data and constraint factors in the system, while m1, m2, and m3 and y1, y2 and y3 are the momentum and deformity variables respectively. K2 are the constants of the spring. The function "k" is constant if the ground effects are not simulated, and non-linear if the ground is soft.

Onesideness is considered by first having the pipe pass through all the defined points, then finding the point of maximum negative reaction (absolute value, which is to say the force pushing the pipe down toward the ground, and releasing the relative support, only to continue onward until there are no other negative reactions left.

To speed up the calculation as much as possible, a solution using a band matrix is essential.

In cases where the ground effect is simulated by making use of a monolateral, non-linear spring, the solution can be found by following a method of the Newton-Raphson type.

The program allows for the introduction of weights and rigidity into each element, horizontal and vertical loads into each node, the application of monolateral, bilateral, linear, and non-linear springs, as well as the introduction of spring buoys and the analysis of the contact between the pipe and the ground.

The Prefabrication Site

A prospective site was picked out for the prefabrication and assembly of the strings (see fig. 3).

The requirements were as follows:

- sufficient length (at least 500 meters);
- respect for environmental restrictions;
- a sufficient area of water in front of the site to allow for the mooring of the line during construction (at least 5 Km);
- adequate shelter from the wind;
- nearness to the installation sites.

The site was found near Colico, or, more specifically, at the mouth of the Adda River where it empties into Lake Como. The yard was, therefore, set up in this uninhabited and uncultivated area formed by the material brought down by the river and deposited parallel to it.

The space available in the hinterland exceeded 500 meters, but there were still obstacles to be overcome such as the fact that, since the area

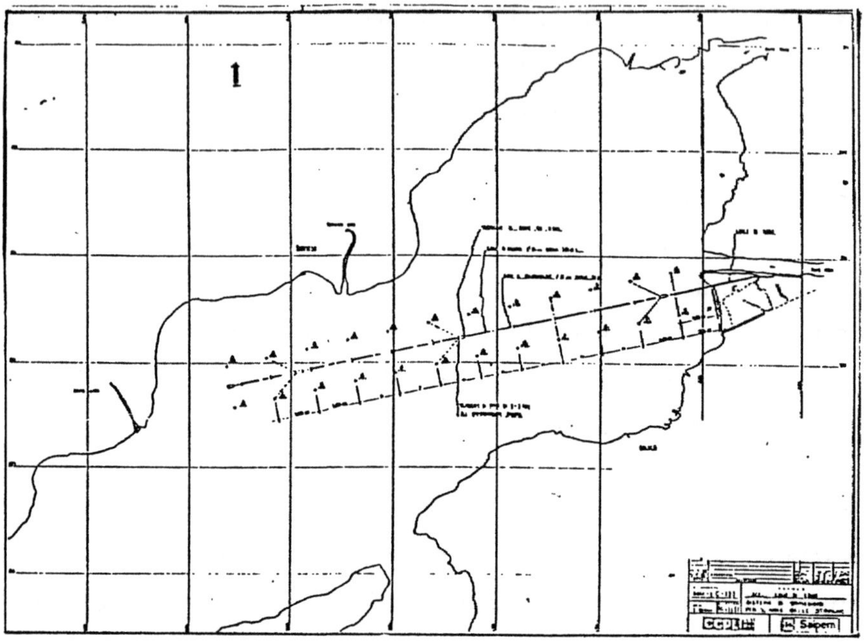

Fig. 3 - Mooring System in Prefabrication Yard

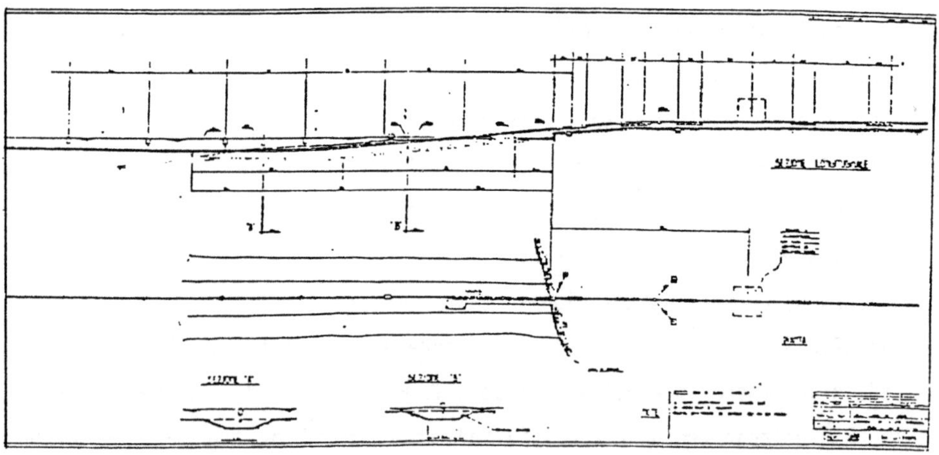

Fig. 4 - Prefabrication Yard Firing Line

flanked the mouth of the Adda, the lake floor sloped downward at a very leisurely pace.

Therefore, an excavation had to be made so that the pipe, once launched, would not touch bottom, since the floats, connected by a 1 meter cable, would sink down by about 70 cm (see figs. 4 and 5).

The excavation was, therefore, made to guarantee a depth of at least 2 meters to the yard's shoreline.

Another problem was that the prevailing winds blow in at an angle of about 30° to the axis of the pipe. Consequently, the idea was developed of placing buoys on both sides of the line every 400 meters for mooring both the tug used to to haul the line and the line itself (see fig. 3).

The distance of 400 meters was set upon as a result of mathematical simulations made on the basis available meteorological data.

The string therefore had a 400 HP tug moored up ahead of it with cables to the buoys lying furthest out and with its engines running at the same time to guarantee a constant pull on the line. Every 400 meters two cables connected an equal number of buoys to the line (see fig. 7), while, on shore a stabilizing winch held the line in place. In fact, to minimize stress during a storm, it was necessary to make sure that there would be a pull of at least 3 tons on the string at all times.

It should be noted that during the preparation of the second string, a severe storm actually did break and, although it managed to damage several float clamps disabling their release mechanisms, with the result that many floats got unhooked from the line, it in no way compromised the integrity of the line.

In sum, the 400 meter strings were prefabricated in the yard, unmoored 400 meters at a time, moored once again and towed by a winch located aboard the tug.

The lateral stability of the strings was therefore guaranteed by the mooring system, the tug and the stabilizing winch on shore.

The clamps were made ready and hooked to the buoys as the 400 meter strings were gradually towed outward by the winch on the tug, itself moored to the two buoys furthest from the shore.

To activate the clamps a cable, with a diameter of 5.5 mm, had to be inserted into their moveable hinges and fixed with a longitudinally halved pin (see fig. 6).

A drum containing the cable was attached to the front part of the string (see fig. 8).

The Floats (fig. 5)

The floats were made of plastic and were designed and tested to withstand the pressure to a depth of 300 meters. They were placed at a distance of 11 meters one from the other and connected to the pipe by a lacing line one meter in length and a spring catch.

The serviceable thrust exercized by each float amounted to 105 Kg.

The Clamps (fig. 6)

The mechanical releasing system devised for the project was based on a series of clamps capable of detaching themselves by pulling on a steel cable 5.5 mm thick fitted with a forced, calibrated head.

With the movement of the cable, the head caused the two-halved pin inserted into the hinge of the clamp to slip out. Once the pin was out, the thrust of the float caused the clamp to open.

A remarkably long series of tests carried out in the workshop led to this simple solution which meets the following requirements:

- reliability;
- speed of assembly;
- the releasing action had to be effected using the means already available on the lake, which is to say with life boats that could give, by their engines, a maximum pull of 30 Kg.

The system worked quite satisfactorily, one of the few drawbacks being that, in bad weather, the motion of the floats tended to weaken its resistance to stress, but this inconvenience was overcome by greatly reinforcing the clamps.

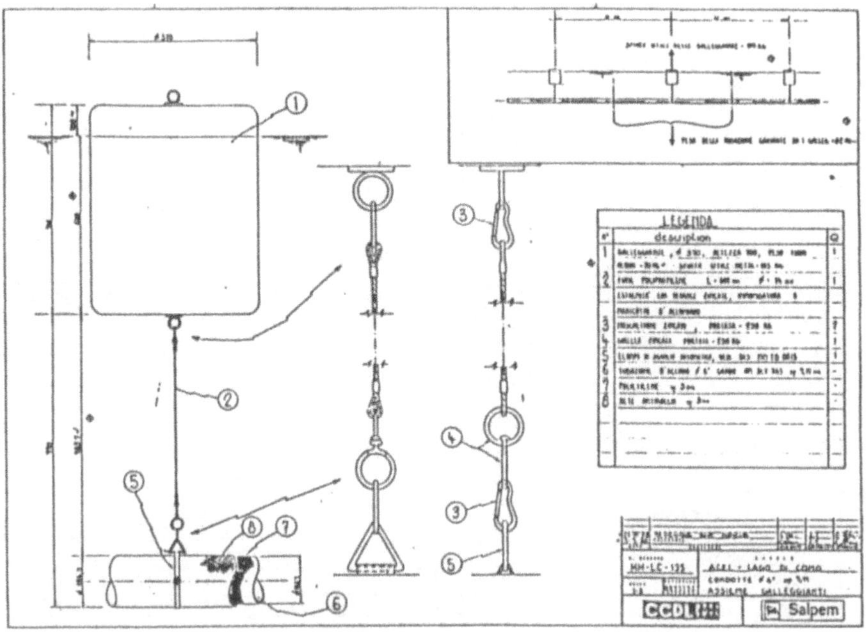

Fig. 5 - Floaters

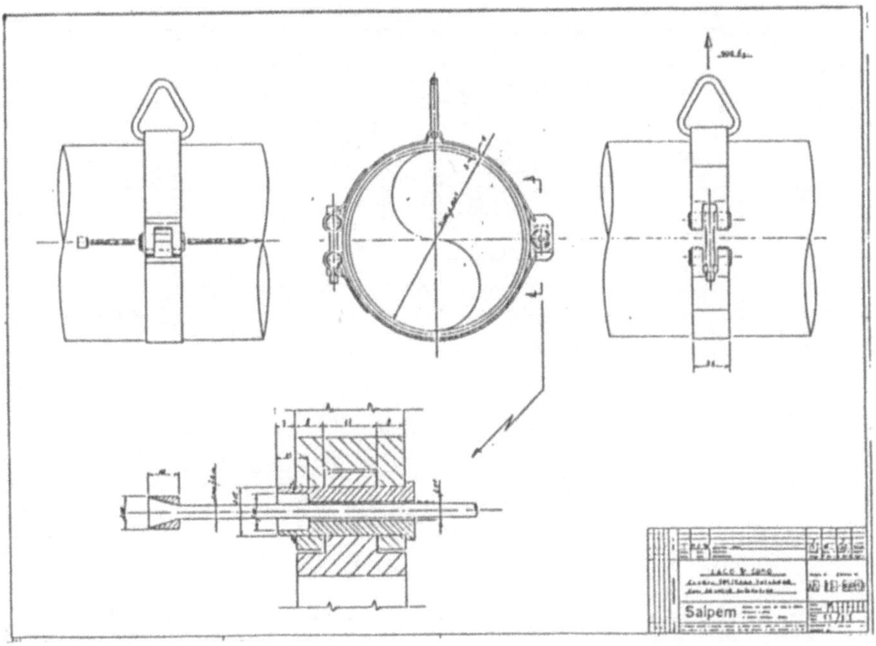

Fig. 6 - Clamps and their Mechanism

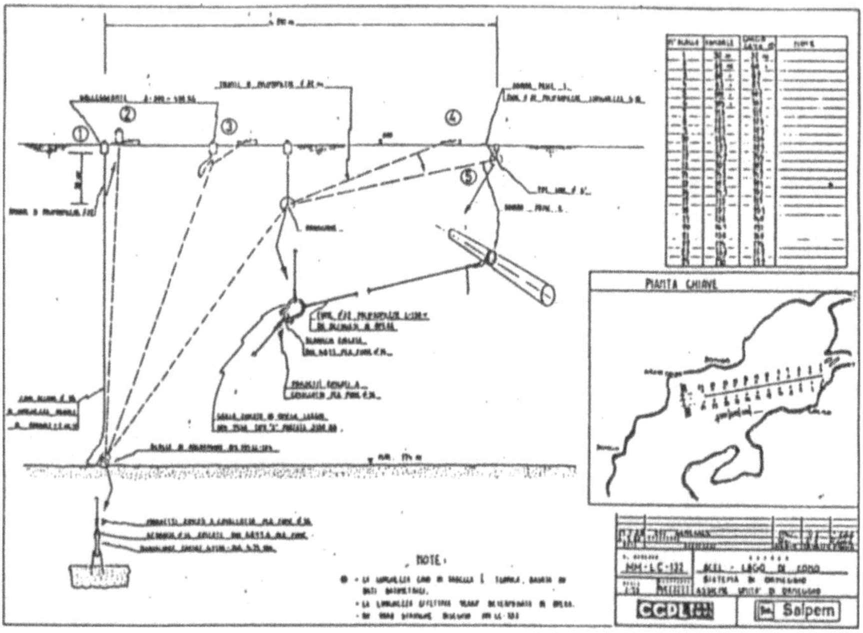

Fig. 7 - Mooring System -
Particular for Prefabrication Yard and
Laying Sites

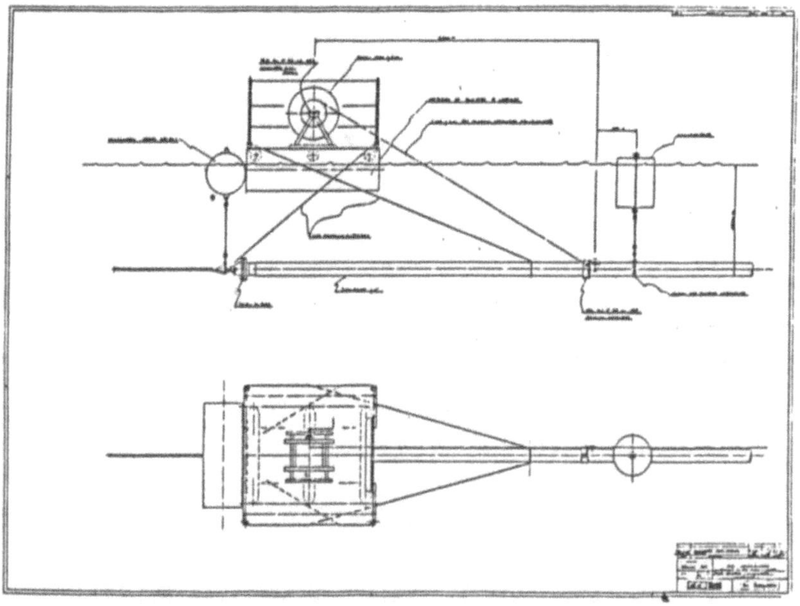

Fig. 8 - Particular of Drum
to pay-out the 5.5 mm Cable

Another inconvenience had to do with the launching operation in the deeper sections of the lake where the release of the floats would occur porthumously owing to the large angles created between the line and the floater's axis during launching, sometimes reaching as much as 15°.
Consequently, the clamp would sometimes get caught in the rock protective netting around the pipe.

The Means

The equipment used throughout the entire installation proceedings consisted of:

- 1 400HP tugboat with a towing winch on board;
- 1 400HP motor-boat;
- 4 lifeboats of our lay-barges.

Navigation

Navigation created no particular difficulties, since, even under adverse conditions, we were able to keep things perfectly under control.
The real problem had to do with the speed of transport in that, even when the tug and motorboat were used together, the speed never exceeded 2.5 knots.
This made it necessary to moor the lines at the laying points to the buoys we had placed there for the purpose (see fig. 7) and leave them there throughout the night.
In fact, even if we started at the first light of dawn, and the lines were close at hand, we never had more than two hours to get through with the preliminaries.
In any case we always managed to sink the line part way, by releasing the largest possible number of floaters, at least insofar as was compatible with the activation of the line stretcher and the difficulties involved in subsequently recovery the floats.

The Tensioning System

In order to guarantee the correct alignment of the pipe with the planned route, it was thought advisable to introduce a tensioning system (see fig. 10). The idea was to exercize a constant force on the string with no geometric restraints.

In fact, from the analyses we conducted, we knew that if the reaction at the point of application did not exceed 650 Kg, the amount of allowable stress would also not be exceeded. If, instead, the line had been subjected to restraints of a geometric nature only, even small errors of measurement might well have led to the development of great amounts of stress at the point of restraint, with no control over the state of tension thus created.

The first system used on the first line, involved the application of 200 Kg counterweights (see fig. 10) which, by means of a pulley mounted under the mooring buoys whose thrust amounted to 1,200 Kg, were made ready, each with its own cable, to be hooked up to the line which had been securely moored during the first stages of the installation.

The system worked well, but turned out to be extremely hard to set up, primarily owing to the fact that the mooring buoys could not be left free to float on the surface, where they would have interfered with the navigation on the lake, but had to be anchored down 5 meters below the surface and their locations marked off only by small, plastic signal buoys placed there some hours before the arrival of the string with its mooring cables. Consequently, in order to attach the counterweights divers had to be employed to work on each mooring buoy.

Moreover, the 200 Kg weights did not make for easy handling and manoeuvring, and their cables tended to get tangled up in the pulleys.

To conclude, while it took only one hour to moor the first string, the installation of of the entire system took several days, thus exposing the line to environmental impacts which could have damaged the float release system. The stretcher cables were afterwards cut with the ROV.

The up shot of all this was that for the installation of the next two lines, the line-stretcher system was changed entirely (see fig. 11),

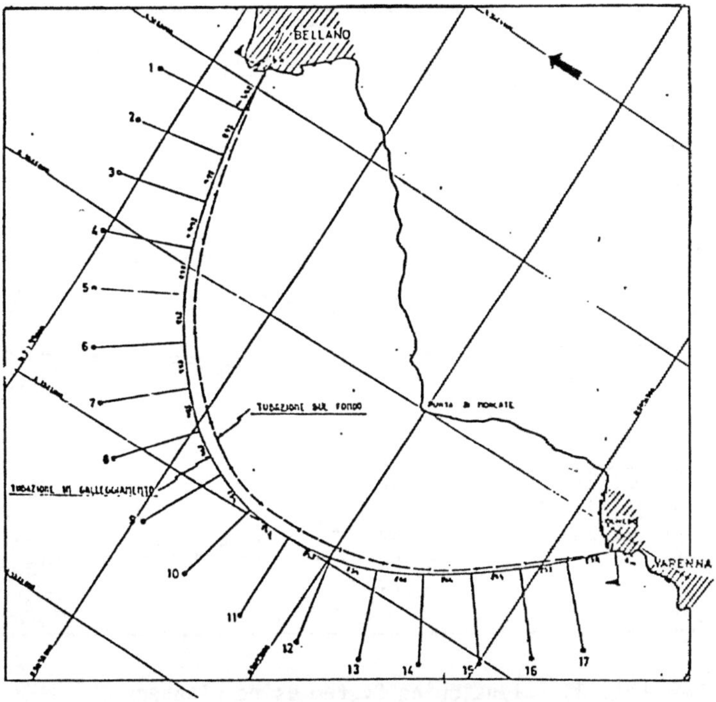

Fig. 9 - Route and Tensioning System of Line Bellano Varenna

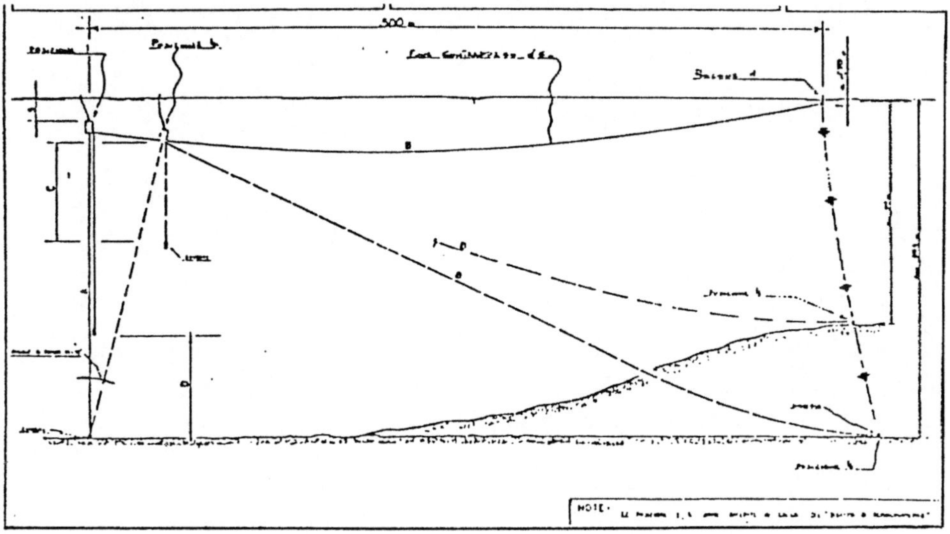

Fig. 10 - Counterweight - Tensioning System Schematic used in First Line

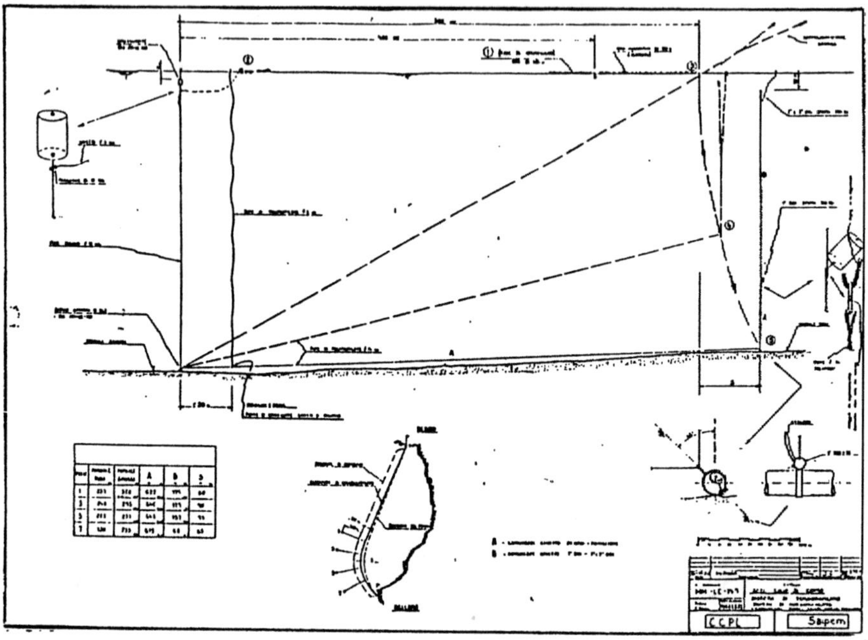

Fig. 11 - Tensioning System using Floaters for Second and Third Lines without, of course, undermining the basic concept of applying a constant force on the line while it was being lowered into place. This time around, groups of three buoys were used so that at each point of restraint a thrust of 300 Kg would be obtained.

The desired effect was obtained by fixing a previously prepared cable of the correct length to the mooring buoy rope, running it through rings attached to the pipe and attaching to its free end another ring to which the group of three floaters could be tied.

With respect to the previous system whereby the force was exercized by the counterweights immediately upon their attachment, with this new method, the forces tenting to stretch the line out were applied gradually, the three buoys being dragged down with it as it was lowered.

The advantage was that the same cables used for temporary mooring purposes could also be used for the stretcher system, without having recourse to divers during both the setting-up and dismantling stages.

More important still was the fact that the entire operation could be handled using material already on hand. Furthermore, the time required to install the system dropped from several days to just a few hours.

The release and Recovery of the Floaters

One of the operational problems that had to be faced was how to release and recover the floaters of which more that 400 were used in getting the last string into place.

It was primarily a question of making the line safe in the shortest possible time which involved the release of the largest possible number of floaters during the few hours available to us.

The drawback was that, owing to fact that navigation was necessarily slow and it was in winter with very few daylight hours (from 1 to 3 depending on the particular line involved) available to carry out the preliminary mooring operation and release and recover the floaters.

Therefore, once the line was moored, at one sitting the release of the floaters could never be total, since we had to make sure that enough daylight was left to allow for the recovery of the floaters.

The ROV

The ROV turned out to be extremely useful in:

1) carrying out a preliminary and accurate batimetric campaign and analysis of the lake bottom;
2) cutting and releasing the counterweights of the line stretcher from the first line;
3) cutting and disengaging those floaters which, for one reason or another, did not come loose from the line;
4) as a back-up system to the mechanical system used to disengage the the floaters;
5) conducting the final survey of the line as laid.

The ROV was a Spring 600 equipped with a Ferranti Ore Trackpoint System II to perform the positioning operation.

Testing

All the strings were pretested in the prefabrication yard by means of a hydrotest during which they were subjected to a pressure of 50 bars for four hours.

Once installed, the lines were again tested using nitrogen at 50 bars.

Conclusion

This job was crowned with success from a technical as well as an economic standpoint, and was completed in record time.

It proved again that, even under the most extreme conditions (depths of more 300 meters), excellent results can be obtained using the simplest equipment with respect to the great lay-barges used to lay pipelines in the sea, by adapting to the particular environment in which the job is to be done, and, above all, by carrying out an accurate preliminary simulation study as the principal tool throughout.

HAZARD AND PROTECTION CONCEPTS FOR DEEPWATER PIPELINES: THE ENVIRONMENTAL FACTORS.

Julio E. MELEGARI , SAIPEM

Giorgio BRESSAN , SAIPEM

San Donato Milanese, Milan (Italy)

Pipeline stability in critical areas of unstable and/or rapidly evolving morphology is a problem of paramount importance not to be overlooked in every pipeline project. The paper deals with the merging aspects of environmental sciences and technical engineering for pipeline stability.

Volume 24: C. P. Ellinas (ed.), Advances in Subsea Pipeline Engineering and Technology, 321–353.

Introduction

Various environmental and natural hazards with potential risks of substantial damage to submarine pipelines may exist along a selected pipeline route.
A considerable number of pipeline damages have occurred because of unpredicted and sudden changes of the original morphology of the sea floor on which they rest. The evolving shape of the sea floor is dependent not only on the local geologic setting and nature of the substrate, as generally thought, but it is strongly influenced by the dynamics of the overlying water mass that makes each marine basin a complex dynamic system. Hydrodynamic conditions of submarine environments are very important factors responsible for the endless modification of sea floor morphology, since submarine currents produce erosion, transport and deposition of sediment.

1. Pipeline survey and hazard identification

To identify the hazards which may exist along a proposed pipeline ruote, data must be gathered regarding waves, surface and subsurface currents, and other data.

Environmental and natural hazards can be classified into two main categories: hazards which pre-exist and can be encountered during the construction (i.e. pipeline installation) on the seabed and during its planned operational life. The specific hazards and severity of the of these hazards depend on the pipeline-site location, whilst the protection works to be performed and the corrective actions to be taken depend mainly on the water depth and on the type of hazard. In gulf and delta areas the pipeline may be exposed to mud slides and turbidity currents as well as to potential severe storm consequences induced underwater and other major bottom instabilities. In the near shore areas, the pipeline is normally exposed to high hydrodynamic forces and actions if it is just installed on the seabed without any protective trenching. In other areas, depending upon local situations and conditions, pipelines may have to be designed and installed considering earthquakes as well as active faulting which may occur in the area. When installed across straits, channels and narrow seas communicating with larger basins, pipelines may be subject to strong bottom currents, migration and collapse of dunes and sand waves, slumps and other slope instabilities. When laid across wide areas of shallow or relatively shallow seas, the pipeline is very likely to rest across a sequence of aggradation bedforms and sedimentary rhythmic mounded obstruction and parallel depressions which may cause unsupported free spans to the possible extent. Sand waves, dunes and other large, current generated bedforms with a ripple shape may become a major problem in the stability and protection of pipelines crossing wide seabed areas where these structures are present and subject to current induced migration. One of the most critical and still uncertain questions to geotechnical and costruction engineers is the stability of these larger bedforms in relationship with the pipelines and the related offshore installation they support. Although the actual size of these sand bedforms may not form an important direct threat to piling of vertical offshore structures, the turbulence and detachment of flow can induce vibrations on tubular components. If the resonance in different pilings varies, damaging stresses can occur in pipelines and, in the case of fixed structures, in horizontal beam connections at the lower levels. Scouring normally is not considered important, as a structural damage, for vertical installations if pilings are set deep enough into underlaying formations. Scour affects large structures in two ways. Global scour which occurs around the whole structure and local scour which occurs around the legs, spool pieces and bracings forming smaller, deeper scour pits. The resulting erosion of material around the jacket base increases, undermining the structural integrity of the foundation in the case of gravity structures and of anchoring pilings too shallow.

Based on the current concepts for identifying the various hazards along a proposed route, the basic criteria in selecting pipeline routes, particularly on unstable seabottom, include special environmental considerations and surveys. Whether developing a deep water prospect or laying pipelines across deep water zones, detailed geophysical surveys are undertaken to establish if any geological or geotechnical problems exist which could affect operations.

Basic criteria for the selection of the most appropriate and safest pipeline route across unstable areas. also include the following concepts: minimize pipe lentgh in unstable seafloors and route the pipeline in a relatively more stable area. In mud-flow areas, minimize any soil movement and slump risks of damage to the pipeline by routing the pipe in such a way that in runs in the same direction as the ascertained or most probable mud flow. This can normally be accomplished by having the pipeline routed in a direction perpendicular to the bottom depth contours in the area.

2. Support and protection of pipelines

To minimize potential risks of damage to the pipeline, the environmental hazards must first be identified in the specific site, then measures be taken to protect the pipeline from these hazards. The protection methods include trenching the pipeline below the seabed, anchoring of the pipeline, increased concrete coating, installation of supports installation of load/protection mattresses, gravel dumping and strengthening the pipeline. In the choice and the application of the most adequate and effective protection method, water depth plays a relevant role as a determining factor. There are methods suitable for installation and actuation by a manipulator arm operated from inside a submarine or from the surface on a remotely controlled vehicle, whilst there are other methods which require the hyperbaric diver and can be actuated by the direct intervention of the human hand only. Some protection methods, on the other hand, can be put into action in more than one way depending upon the water depth and whether the human hand or a remotely controlled work system is used. Of course the different installation system and procedure has an economical impact basically dependent on whether man-in-the-sea techniques are employed or not.

For stability and protection against mechanical damages, most offshore pipelines rely on the application of concrete coating of variable thickness. The design thickness of the concrete coating depends on factors such as seabed currents, pipeline, size, buoyancy, concrete density, required corrosion protection and resistance to mechanical damage and spanning. Natural irregularities in the seabed topography, geomorphic discontinuities or scour under the pipeline during service normally result in substantial free lengths of unsupported pipeline which can not be tolerated, as che concrete coating becomes overstressed and spans leading to loss of protection and instability and eventally to permanent damages to the pipeline structure. In common practice freespan redustion and correction is achieved by providing intermediate supports, augmented where necessary by additional weight and perimeter protection.

2.1 Sandbags

Sadbagging is carried out to support free spans of cover exposed pipelines. It is very often the simplest and most cost effective method for depths down to 50mt. (Figure 1)
Experience and tests in the North Sea have shown that the conventional sandbags give a rather low level of protection against mechanical damage, while Gulf Coast experience using sand/cement mixtures has been effective.
Return line to the surface in order taht the grout can be recirculated when necessary.

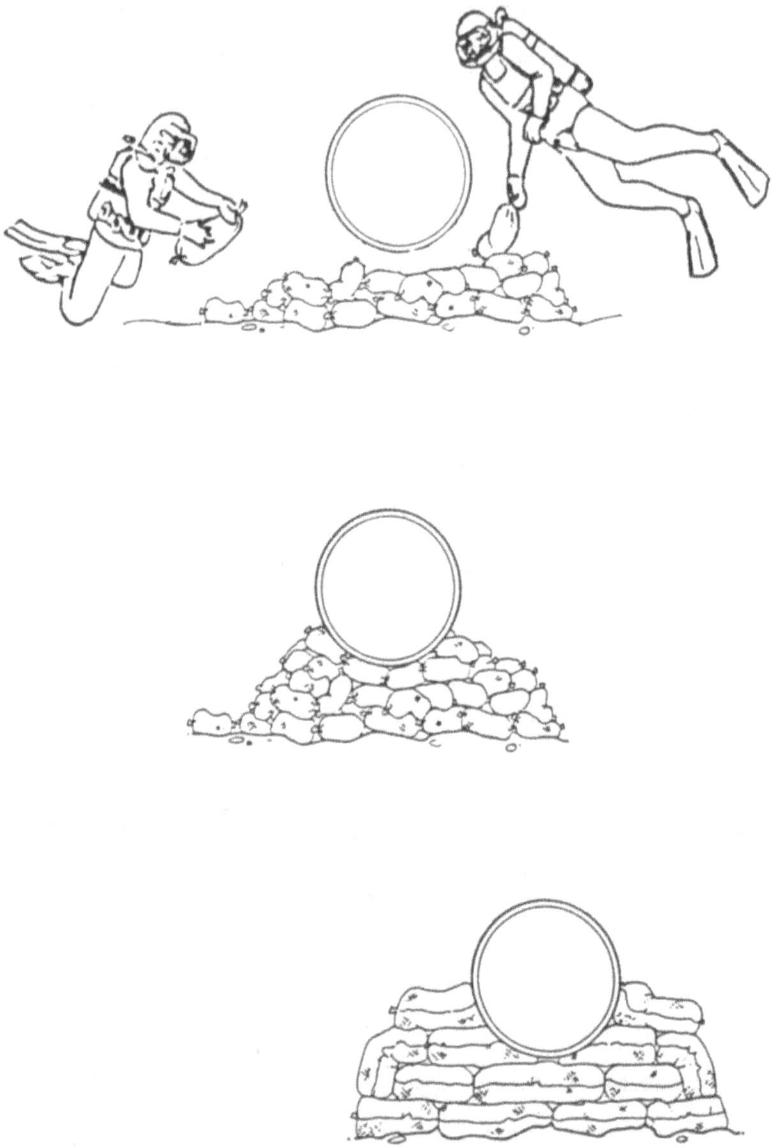

Figure 1 Sandbagging

2.2 Grouting (support)

This engineering solution, to be installed by hyperbaric divers under normal environmental and technical conditions *(shown in Fig 2)* , is based on a well proven flexible fabric which is constructed to form bags or mattresses when filled with cement grout. The formwork is made from a purposewoven polypropylene fabric, the support/underpin form being based on several interconnected compartments with vent pipes to ensure correct filling and maintenance of contact with the underside of the pipeline for a standard distance/clearance. The system is tailored to inflate the pipeline and the bags can be adapted to accommodate varying heights of undercut beneath the pipeline. The filling grout is pumped from the surface support vessel through a grout umbilical consisting of two pressure hoses which provide a return line to the surface in order that the grout can be recirculated when necessary.

2.3 Jack-ups (mechanical supports)

In several cases specially designed active supports have been used to reduce the length of free spans with variable bottom clearance. (Figure 3)

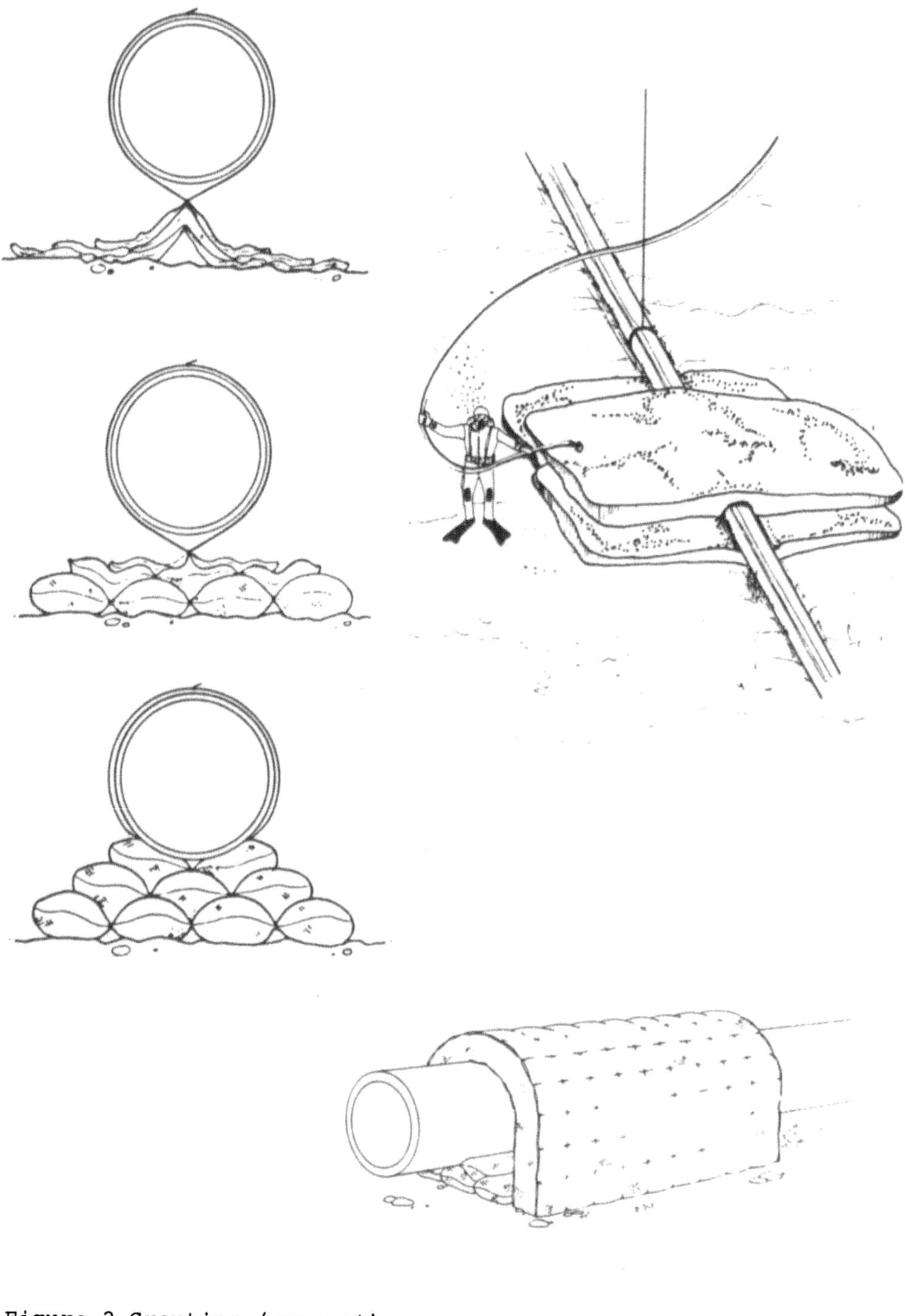

Figure 2 Grouting (support)

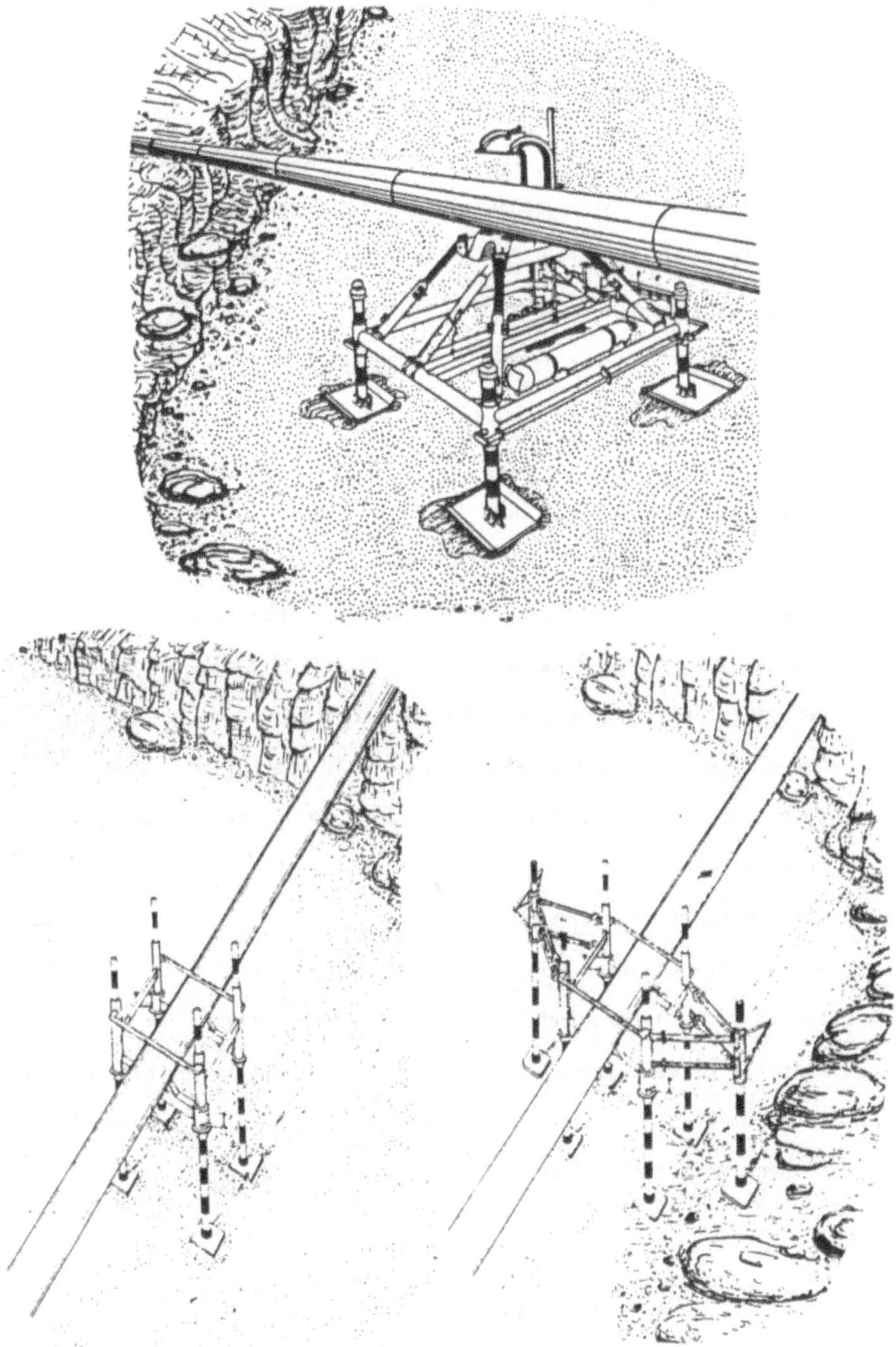

Figure 3 Jack ups (mechanical supports)

The supports can, by means of releasable legs, adapt themselves perfectly to uneveness of the seabed. They are equipped with hydraulic jacks and compressed air cylinders that allow the pipe to be jacked-up so that it has an optimal configuration and stress status. In this case they become "active supports".
These types of supports are today suitable for installation at depths exceeding 500m with the assistance of a D.P. vessel and a manned submarine or a remotely operated vehicle.

2.4 Gravel dumping (backfilling)

This method is widely used for correcting free spans. In this way a free span can be covered along its total length or only at certain points by forming heaps of material.
Gravel backfilling can also be used in combination with pipeline steel supports. The practice is to cover the supports after installation so as to provide more stability and protection from erosion under the supports themselves.
One important factor for the engineered backfilling is to determine the correct mixture and grain size of the backfill material. The backfill material must remain in place during the different environmental conditions that may occur. Also, it must not prevent fishing with bottom towed fishing gears in the area.

For water depths to 50m backfilling by use of side or split dumping barges can be used depending on the current in the actual area. However, owing to the spread of the dumped gravel, placement can be inaccurate and consequently unacceptable. On these vessels, the gravel is carried in compartments at deck level from where hydraulically operated shovels push the dumping material progressively over the side of the vessel resulting in a continous flow. During dumping the vessel can keep station or move along on a predetermined track.

In deeper waters the most efficient way of dumping gravel is through a guide pipe. The lower end of the fall-pipe is kept about 10m above the pipeline, and by the use of conventional navigation positioning control equipment, the vessel maintains its position over the pipeline.

These vessels are preferably dynamically positioned, and equipped with acoustic control devices and navigation system.

2.5 Grouting (protection)

The fabric formwork used for the installation of grouted supports can also be tailored in the form of a saddlebag to provide additional weight coating or protection over the pipeline. In such circumstances the same fabric material is tailored and adapted to suit the pipeline size, height from the seabed and specific weight. (Figure 4)

2.6 Bitumen mattresses

Bitumen mattresses are considered more suitable for the protection/stabilisation of exposed sections of pipelines in deep water. Whenever an unsupported span is of considerable length and its clearance from the seabed is very little (centimetres), its correction can be achieved by increasing the pipe's negative buoyancy (i.e. by pushing the pipeline onto the seabed) provided that it is maintained within the allowable stress limits.
These mattresses can also be used in combination with other methods, whenever negative reactions in the pipeline are required (i.e. support stability, etc.).
Bitumen mattresses, to be fully effective, need to be properly sized in relation to pipeline diameter. They are extremely heavy for their size and with their flexibility, they can provide a very effective cure to difficult stabilization problems. The bituminous filler combined with dense aggregates is used to provide weight, flexibility and long lasting protection to the pipeline.(Figure 5)

2.7 Concrete mattresses

These mattresses are constructed from reinforced concrete bars interconnected by steel or polypropylene ropes which provide flexibility and capability to cope with an uneven seabed profile. This type of protection is suitable for multiple application in remedial works on pipelines (repair of weight coating, mechanical protection, pipeline stabilization, etc.) and in the prevention of scour. (Figure 6)

2.8 Concrete saddles

Concrete saddles can be used instead of sub mattresses to ensureadditional weight coating and protection on the pipeline and to protect it from local mechanical damage. These saddles are suitable for repair of weight coating and local scour prevention. They obviously provide good "mechanical" protection to a pipeline and can be easily installed by divers or remotely controlled vehicles. (Figure 7)

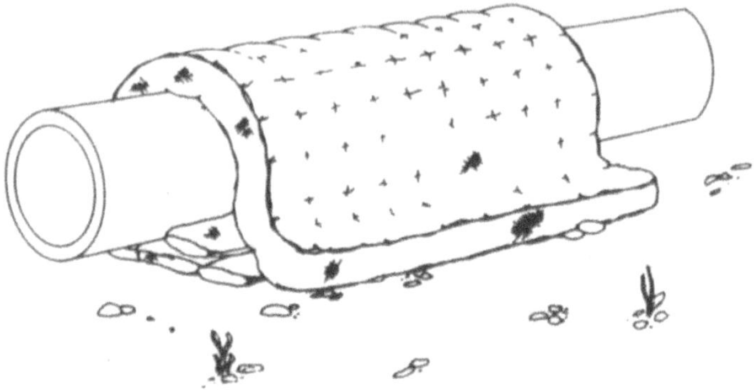

Figure 4 Grouting (protection)

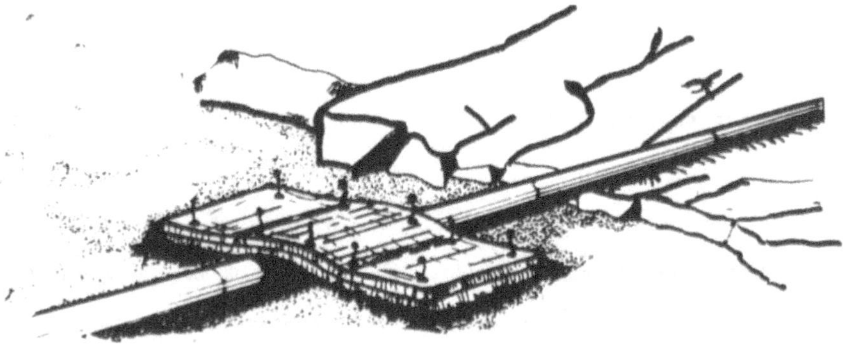

Figure 5 Bitumen mattress

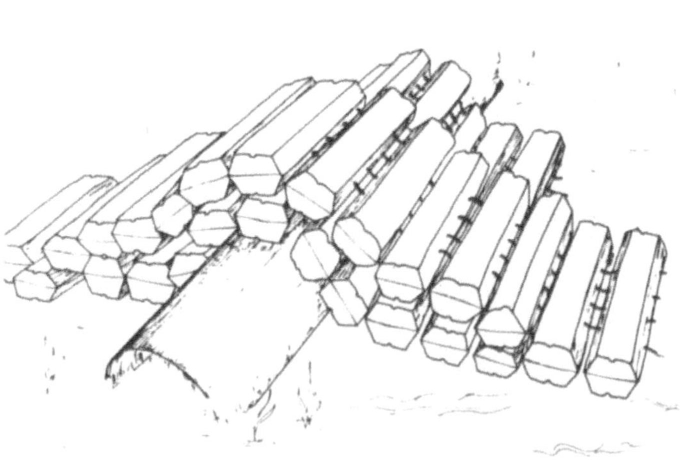

Figure 6 Concrete mattress

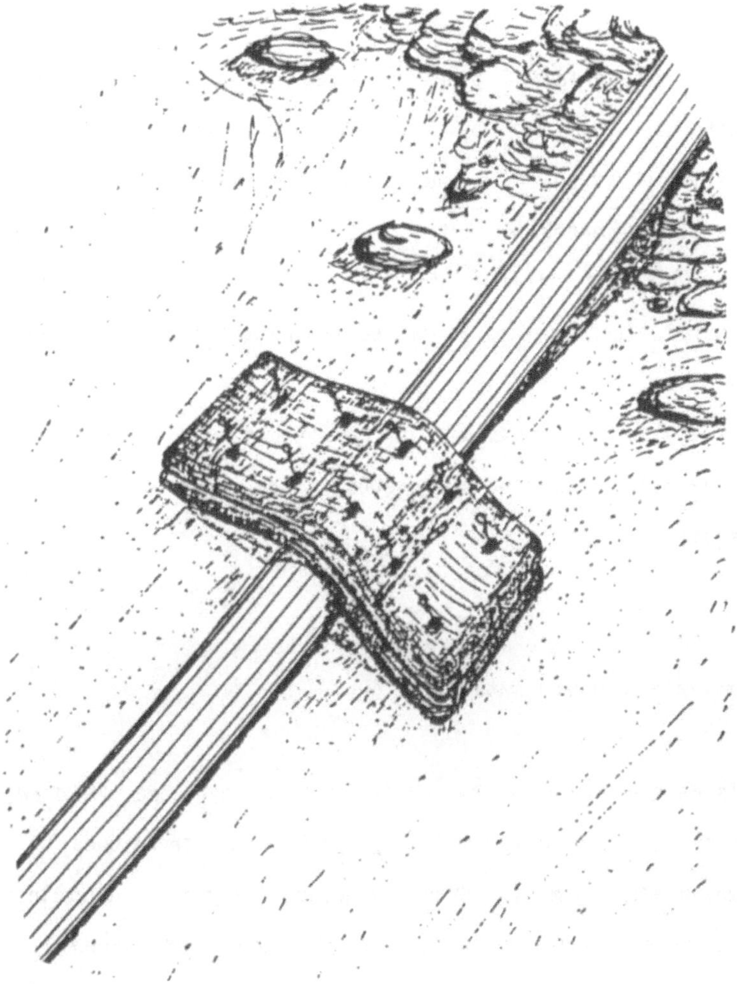

Figure 7 Concrete saddle

2.9 Anchoring systems

In critical areas of pipeline, such as a shore approach or near platforms, it may be necessary to anchor the pipeline by physically fixing it to the seabed to eliminate longitudinal or lateral movements.
Two piles are driven into the soil, one on each side of the pipeline, and a clamp is fitted around the top of the pipe and to the piles.
This system is independent of the seabed soil because the anchors can be piled, drilled or screwed to the depth required to provide adequate restraint.

2.10 Artificial seaweed mats

This method has been devised in order to overcome the drawbacks of the two most commonly used scour protection techniques: gravel dumping and sub mats. Gravel dumping, under certain local conditions, requires further maintenance and control due to settlement of the stones. With sub mats, as with any rigid profile object, edge scour still occurs, resulting in slow settlement and requiring further intervention. (Figure 8)

The fundamental idea of this method is based on a technique which has been used for centuries, whereby dunes and quicksand areas were stabilized by means of grass. Following this principle, artificial seaweed sewn to weighted mats is utilized in areas with a moveable seabed to reduce local flow velocity and turbulences in the vicinity of the pipeline, thus not only preventing erosion, but also causing the building-up of sand between the sinthetic seaweed frond.

This system is based on building stable mass fibre reinforced banks. These banks are created by mats of polypropylene fronds that have a dual action. The apply viscous drag which reduces the current velocity so that particles of sands are deposited into the mat - thereby building up a cohesive underwater snadbank around the structure to be protected.

Experience from various locations and environmental conditions shows that this equipment is not always fully effective and the reasons will not be fully known until more research into seabed scouring has been carried out.

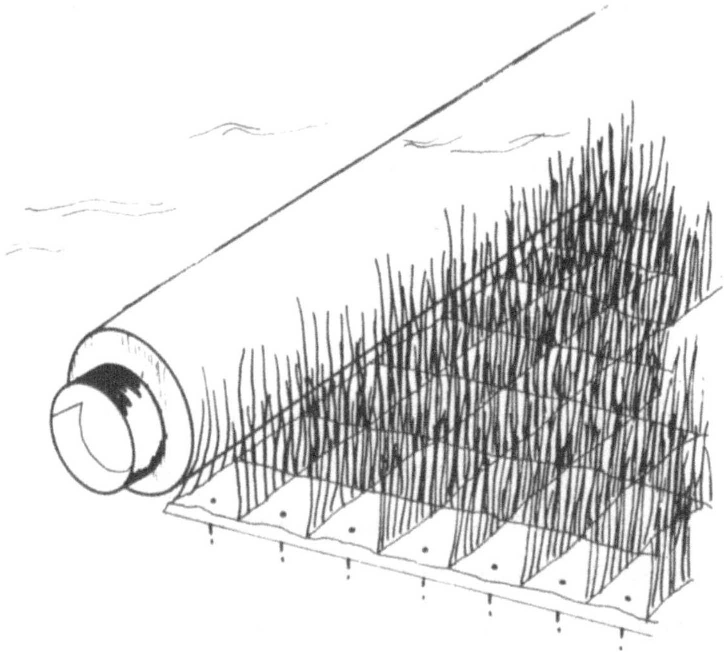

Figure 8 Artificial seaweed mat

3. Physical processes responsible of high hydrodynamism in shallow and deep sea

High-energy conditions of the water masses are common in shallow seas. They lie between those parts of the sea dominated by nearshore processes and those dominated by oceanic processes, at depths between 10m and 200m. Shallow seas include continental shelves and partly enclosed basins such as the North Sea, Yellow Sea and Bering Sea. High-energy conditions in shallow seas arise from the interplay of waves and currents. Four main types of currents exist: a) oceanic circulation currents, b) tidal currents, c) meteorological currents, and d) density currents.

Until the 1945 it was common the notion that high-energy conditions were restricted to shallow waters, and that deep seas beyond the shelf edge were stagnant, moving only at the surface in response to wind stresses. The systematic study of deep-sea sediments by the Deep Sea Drilling Project and by the Ocean Drilling Program, shows now as deep seas are dynamic systems of great complexity in which surface currents coexist with deep currents of some strength that are able to substantially reshape the morphology of the sea floor. High hydrodynamism in deep waters is provided by the occurrence of normal bottom currents and by the resedimentation processes triggered by slope failure, such as turbidity currents, debris flows, slumps, etc. All of these processes are capable of eroding, transporting and depositing sediment, giving rise to the development in deep waters of a series of morphologic features, such as channels, scours, flutes, bedforms, slump scars, etc., that represent potential dangerous features for pipeline stability.

Normal bottom currents include all those deep currents that are not driven by sediment suspension, and may therefore flow alongslope, downslope and upslope. They comprise:

- Internal waves and tides. These are large-scale oscillations at density discontinuities between water layers in the upper few hundreds of metres, notably at the thermocline (Lafond, 1962). Internal waves and tides can cause erosion and migration of bedforms at the shelf break, on top of seamounts, or in relatively shallow slope and shelf basins.
- Contour currents. Deep ocean contour currents are formed by the cooling and sinking of surface waters at high latitudes and the deep slow thermocline circulation of these polar masses throughout the world's oceans (Neuman, 1968). Whereas much of the deep sea floor is swept by very slow currents (2 cm/s), some currents can be greater than 100 cm/s where the flow is restricted throughout narrow passages, such as straits. The effects of contour currents on the morphology of deep sea floors include the erosion of channels, moats and furrows, resuspension and transport of fine-grained sediment and the migration of bedforms.

4. Sedimentary processes and features hazardous for pipeline stability.

The main sedimentary processes responsible of morphologic changes on the sea floor that are potentially dangerous for pipeline stability include: slope failure, bedform aggradation, migration and scouring.

4.1 Slope failure

Slope failure is the primary process that alters the morphology of slopes. It includes mass movement of material for which gravity provides the driving force (Dott, 1963). Mass flows are more common on slopes located in areas characterized by high sedimentation rates, such as in deltaic regions (e.g. Mississipi delta, Prior & Coleman, 1980; deltas in Norwegian fjords; Terzaghi, 1956), where the sedimentation rate can reach 3000 cm/y. By the early 1960's the offshore oil and gas industry has experienced a high number of pipeline damages, especially in the upper delta slopes in water depths of less than 30 m. Survey and diver's reports indicate that pipeline are both locally displaced in a downslope direction and appear to have sunk within the sediment. Moreover, mass movements can occur frequently also in deep-water settings and on very low-angle slopes, less than 1°, such as on isolated rises (e.g. Madeira Rise).

The main types of queous slope instability features and processes include: a) growth faults; b) pseudo-diapirs; c) rockfalls; d) slides and slumps; e) collapse depressions; f) sediment gravity flows. Sediment gravity flows can be further subdivided on the basis of their internal mechanical behaviour and dominant sediment support mechanism.

a) Growth faults are faults in sediments along which the movement is contemporaneous to sedimentation. In most instances these faults, which are common in deltaic regions, tend to cut the modern sediment surface, forming abrupt scarps on the sea floor. These scarps can provide localized areas for additional downslope mass movement of material by slumping.

b) Pseudo-diapirs. Pseudo-diapiric intrusions of mud into overlying sands, giving rise to a sudden development of islands in the subaqueous sedimentary surface, occur frequently near the distributor mouths of some deltas (e.g. Mississipi delta, Morgan, 1961; Coleman, 1976). c) Rockfalls. Are rapid accumulations of clasts by freefall. In marine environments they occur at the base of steep slopes, such as carbonate reefs, fault scarps and canyon walls. Displaced clasts may be very large (10m) and bounce or roll downslope up to several hundreds of meters before coming to rest (Abbate, Bortolotti & Passerini, 1970).

d) Slumps and slides involve downslope displacement of a semi-consolidated sediment mass along a basal shear plane while retaining some internal coherence. Slides have a planar shear plane and little internal disturbance. Slumped and slided deposits are very widespread and can range in volume from less that 1 m^3 to over 100 m^3. A large slump generally possesses three morphological components:

a) a source area, or head, characterized by tensional structures, such as faults, slump scars; b) a central transport zone represented by a channel; c) a distal depositional area, or toe, displaying compressional features and overlapping lobes of remolded debris.

A man induced reason for increased concern in areas of potentially unstable sediments on submarine slopes is the post trenching of existing pipelines after laying. Trenching operations generally cause local stress concentrations within the sediments and induce excess pore pressure within the sediments. The results of these stress concentrations and increased pore pressure is often the spreading of submarine slumps. An effect very similar to the consequences generated by pipeline trenching is often produced by anchors and anchor wires during operations offshore carried out by conventional work barges and lay-barges in areas of submarine slopes.

e) Collapse depressions. These features, which occur mainly in shallow deltaic areas (Prio & Coleman, 1980), are bowl-shaped depressions of the sea floor, ranging in size from 50m to 150m. Typically the depressions are bounded by curved scarps up to 3m in height, within which the bottom is depressed and filled with irregular blocks of sediment. These features are the result of liquefaction, and decrease in the volume of the sediment-gas-water system. f) Sediment gravity flows. They occur commonly in deep seas and are associated to slides and slumps. They are divided into: a) debris flows; b) grain flows; c) liquefied and fluidized flows; d) turbidity currents.

f.a) Debris flows are higly concentrated, higly viscous sediment dispersions that possess yield strength and plastic flow behaviour (Hampton, 1972). They are slurry-like or glacier-like laminar flows, advance downslope continuously or intermittently, and the front shows a scarp up to 30m or more in height. f.b) Grain flows are quasi-visco-elastic flows of sand and gravel characterized by grain-to-grain collision. They require slopes steeper than 18° and represent a very localized process in the subaqueous environment. They are frequent on the steep sandy and gravelly deltaic slopes of Gilbert-type deltas (Colella et al. 1987).

f.c) Liquefied and fluidized flows include the collapse of loosely packed silts and sands by upwards moving pore-fluid. Fluidized sand behaves like a fluid of high viscosity and can flow rapidly down slopes in excess of 2°-3°.

f.d) Turbidity currents are catastrophic flows that can reach velocities of up to 100 km/h, where the sediment is kept in suspension by fluid turbulence. They are the most common sediment gravity flows occurring in the deep sea, and can be extremely dangerous for stability of pipelines, since they produce sudden erosion and deposition of sediment. Turbidity currents have been responsible for several breakages of submarine cables. The classic example is the Grand Banks earthquake of 1929 that triggered an enormous slump and an ensuing turbidity current that travelled downslope from hundreds of kilometers on to the Sohm Abyssal Plain (Heezen & Ewing, 1952). The maximum velocity attained by this current was some 70 km/h. Such currents can be up to several kilometers wide and several hundreds of meters thick (Komar, 1969) and can travel as fas as 4000-5000 km (Curray and Moore, 1971).
The frequency with which turbidity currents are generated depends on the nature of the area from which the currents originate, the proximity of the source area, the seismicity of the source area. Turbidity currents generated in deltaic areas after periods of high river discharge may occur as often as once every two years (Heezen & Hollister, 1971). Areas located at the base of slopes, that is in proximity of the source areas of turbidity currents, can be affected by such flows once every 10 years (Gorsline and Emery, 1959).

4.2 Bedform aggradation and migration

Bedforms include undulations of the subaqueous sandy and gravelly sedimentary surface, produced by tractive currents. They can be symmetric when produced by wave motion, and asymmetric when produced by unidirectional and tidal currents. Bedforms include, among others: tidal sand ridges, sediment waves, sand waves.
Tidal and ridges are a very prominent feature of the southern North Sea. They are large-scale linear ridges parallel to the direction of tidal currents. They are up to 40 m high, 2 km wide, 60 km long and have spacings between 5 km and 12 km. The ridges are asimmetric with the steep face inclined at a maximum of about 6°.
Sediment waves are asymmetric undulations of a muddy sedimentary surface with amplitudes of dozens of metres and heights of several metres. They can occur on abyssal plains and continental rise and slopes and have frequently been attributed to the action of contour currents (Damuth, 1980). Some authors have suggested a turbidity-currents origin for sediment waves on channel levees of deep-sea fans (Damuth, 1979; Normark et al., 1980).

Sand waves and dunes represent two big groups of large-scale asymmetric bedforms. These ones are common in the continental shelf at shallow water depths, where they are produced by powerful tidal currents, strong unidirectional flows, surface waves and meteorologically-driven currents (Swift & Ludwick, 1976; Flemming,

1981). However, they can occur also in deep seas (e.g. Messina Strait), where high-energy conditions occur. Because of the confusion existing in the literature about the criteria that distinguish these two types of bedforms, for brevity we will group them in the class of sandwaves.

Sand waves are more or less regularly spaced undulations of a sedimentary surface, and represent the product of the movement of grains under the effects of unidirectional and tidal currents. When these currents are active, sand waves migrate in the direction of the currents. Sandwave wavelength varies generally from 0.6m to many hundreds of meters and the height commonly ranges from 0.05m to 15m or more. Sand waves can have straight or sinuous, continuous or discontinuous crestlines, and occur in groups that form fields of varying shape and size.

Migrating sand waves can be responsible for a rapid evolution of the morphology of the sea floor, and can be very critical for pipeline stability. Potential risks for pipelines crossing fields of active sand waves are:

1) development of free spans the length of which is dependent on the bedform wavelength.

2) vibrations of pipelines due to occurrence of macroturbulence on the lee side of sand waves.

3) collapse of sand waves. This process has been documented by a sedimentologic study of ancient sand wave deposits outcropping along the margins of the Messina Strait (Montenat & Barrier, 1980), which are similar to the sand waves that are presently migrating on the sea floor of the strait.

4.3 Scouring

Erosion results in truncation of the sea floor and development of depressions of varying shape and size that can be potentially risky for pipelines, because of the development of free spans. Erosional features, of whatever origin, occur over a wide range of scales, up to hundred of metres deep and kilometres wide, and over a wide range of water depths.

There are two main processes that produce erosion.

a) erosion by mass movement downslope, creating a feature commonly arcuate along the slope, i.e. a slump scar. This type of structure has been already described in the slope failure section.

b) erosion by water scour, creating generally features elongated in the direction of fluid movement, e.g. channels, chutes, flutes, etc.

A third type of process,which is not widespread as the previous ones, but can create hazardous conditions for pipelines located at high latitudes is represented by ice scouring. The fact that large,

deep draft icebergs may contact the seabed raises a number of concerns in the design and the protection of seabed structures such as pipelines and related installations (wellheads, etc.). The interaction between the iceberg keel and the seabed slope can be modelled as a ploughing phaenomenon in which the sediment in front of the iceberg keel is heaved up in a series of passive failure surfaces and displaced to the sides as the iceberg keeps moving forward under the action of current and wind. In general the resulting scour resembles a long linear or irregular trench or furrow of gradually increasing depth with berms on both sides.
Critical conditions for pipeline stability arise mainly by the sudden and unpredicted development of erosional features on an originally flat surface. Scouring can be triggered by small irregularities of the sea floor of biogenic (e.g. fish scouring) and sedimentary origin (e.g. sand volcanoes). Small bumps or depressions on the sedimentary surface will cause acceleration of the flow which gives rise to flow separation. The associated higher shear stresses lead to erosion which, in turn, emphasises the relief near the irregularity, and so on.

Fish scouring

Local scouring can be triggered and/or caused by the presence of fishes permanently living in proximity of the seabed and in close proximity of natural and artificial structures. Fish scouring can lead to the formation of hollows generated by the movement of the fins (active fish scouring) in search of food near a pipeline or for the excavation of a more protected shelter. In some environmental conditions the mere presence of the fish in close proximity of the seabed is sufficient to induce scouring effects depending upon the current speed (passive fish scouring).

References

1. BARRIER Pascal, 1986 - Evolution paléogéo-graphique du dētroit de messine su Pliocène et au Pléistocène - Giornale di Geologia, 48, 7-24

2. BENEDINI A., MELEGARI J.E., 1989 - Ten years of experience in deepwater pipeline tecnology and maintenance : Strait of Messina 1979-1989" - DOT Deep Offshore Tecnology Conference - Marbella

3. COLELLA A., 1988, - Braid deltas, shelf-type and Gilbert-type fan deltas in the Crati Basin (Pliocene-Holocene, souther Italy): A consequence of varying tectonic and structural conditions - In Nemec, W., and Steel, R., eds. Fan deltas

4. COLELLA A., 1988 - Marine fault-controlled Gilbert-type fan deltas - Geology

5. COLOS, 1983 - Fabric Formworks and Grouting Offshore - Catalog and Operation Manual - Aberdeen

6. Dott. R.H., 1963 - Dynamics of subaqueous gravity depositional processes - Bull. Am. Ass. petrol. Geol. 47, 104-128.

7. EMBLEY R.W., 1980 - The role of mass transport in the distribution and character of deep-ocean sediments with special refernce to the north Atlantic - Marine Geology, 38, 23-50

8. HAMILTON D. & SMITH A.J., 1972 - The origin and sedimentary history of the Hurd Deep English Channel, with additional notes on other deeps in the western English Channel - Mem. Bur. Rech. geol. minieres, 79, 59-78

DEEP WATER TIE-IN

PAAL PETTERSEN, TOMMY BJØRNSEN AND OLE T. MYKLESTAD

AKER ENGINEERING BERGEN A/S
(Previous Bergen Engineering A/S)
Sandslimarka 51
5049 Sandsli
NORWAY

ABSTRACT

This paper, named "Deep Water Tie-In", describes a tie-in concept for large diameter pipelines. The concept is developed for use at large water depths where diving is both costly and dangerous.

The concept is developed for a typical Gravity Base (GBS) platform situated in 300 metres water depth.

The tie-in of pipelines with a diameter of up to 40" is achieved through a tie-in arrangement system pre-installed in the platform substructure. The main components are an atmospheric tie-in tunnel, seal tube and J-tube for pipeline pull cable. The paper describes each of these components in more detail before it embarks on a step by step account of the pipeline tie-in procedure.

In order to establish design loads on the sealtube and identify the optimum pipeline approach, computer analyses have been carried out. It is shown that a curved pipeline approach or a dogleg are both feasible from a load on anchor or stress in pipeline point of view. However, a straight pipeline approach causes loads and stresses too high for practical design purposes.

In order to evaluate the advantages of the proposed tie-in arrangement, the paper ends with an outline of alternative tie-in concepts.

Volume 24: C. P. Ellinas (ed.), Advances in Subsea Pipeline Engineering and Technology, 355–372.

INTRODUCTION

This paper describes a deep water tie-in concept suitable for large diameter pipelines. The concept has been developed by Aker Engineering Bergen A/S (Previous Bergen Engineering A/S) for a typical Gravity Base Structure (GBS) situated in 300 metres water depth. Aker Engineering Bergen A/S has during the last few years performed several studies with regards to the tie-in of pipelines at deep waters.

The tie-in of pipelines up to 40" diameter is achieved through a tie-in arrangement system pre-installed in the platform substructure. This so that the tie-in of the pipelines to the platform riser systems can be achieved in a safe and practical manner with emphasis on flexibility and back-up possibilities.

The tie-in of the large diameter pipelines are to be performed within a dry tie-in chamber through a seal tube system constructed as an integrated part of the GBS structure. This tie-in arrangement has the following main advantages compared to alternative methods:

- The entire riser is installed within the riser shaft and protected against environmental loads and dropped objects.
- Straight forward inspection and maintenance of the riser and corrosion protection system performed by the platform crew.
- Pipeline or pipeline section pull-in directly from a lay barge.
- Tie-in weld is performed in dry one-atmospheric condition.
- No diving required. Pull-in will be assisted by ROV.
- Flexibility with regards to sizing of future pipelines. The pre-installed tie-in arrangement can accommodate a range in pipeline sizes.
- Possible to reverse tie-in in order to change pipelines.

In order to evaluate the advantages of the proposed tie-in arrangement, alternative tie-in concepts have been outlined.

SYSTEM DESCRIPTION

The following is a description of the main systems in the proposed design. Figure 1 shows a typical outfitting of a GBS shaft.

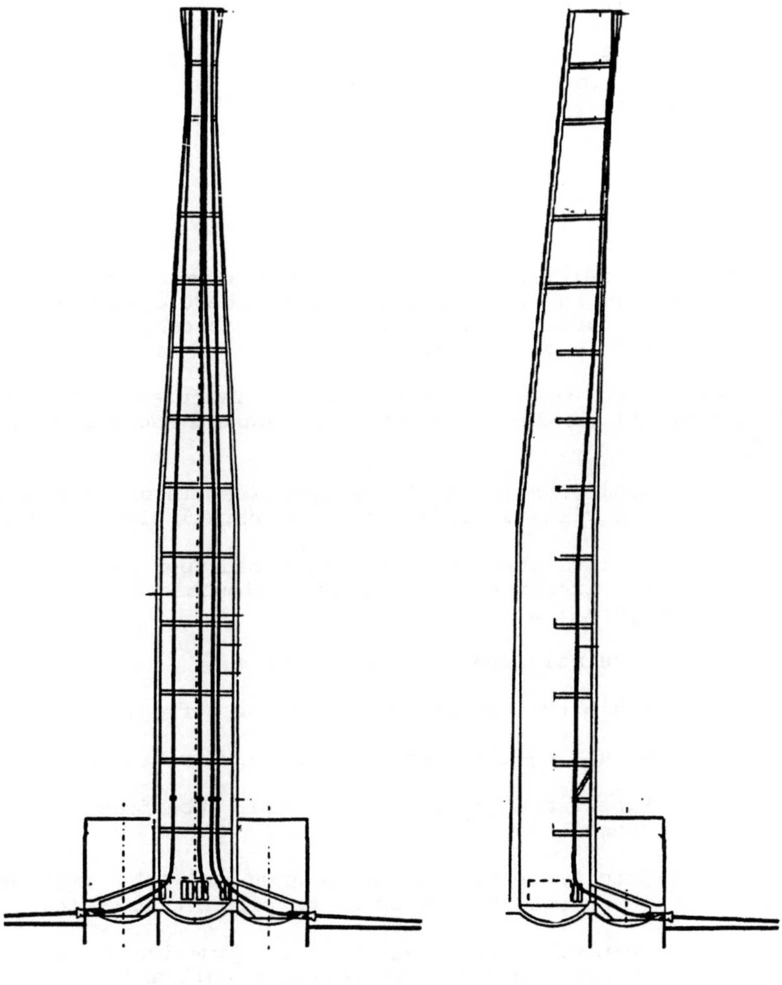

Figure 1 Shaft Outfitting

Atmospheric Tie-In Tunnel

Concrete tie-in tunnels are proposed to be constructed in order to accommodate one-atmospheric tie-in of the large diameter pipelines.

The tunnels are designed to withstand the external pressure and allow direct access from the riser shaft.

The size of the tunnels at the tie-in area is based upon the required size of the seal tubes. It must have sufficient space to lay down materials and perform the tie-in tasks.

It is possible to design a sealed entrance to the tunnel in order to achieve additional safety in the case of sealing failure in the seal tube.

Seal Tube

Each tie-in tunnel will be provided with a seal tube in order to accommodate pull-in of the pipeline into the tie-in tunnel. Based on safety considerations it is proposed that only one sealtube is to be installed in each tie-in tunnel.

The following are some of the main design features of the sealtube with regards to tie-in and operation. Figure 2 shows the seal tube layout.

- Seal tube length to be optimized in order to minimize the risk of jamming due to installation tolerances.
- Inclination and elevation based upon laying operation requirements and pipeline allowable free span during operation.
- Internal pipeline centralizers.
- Inhibitor injection piping for corrosion protection
- Grout injection piping for back-up sealing
- Pipeline anchoring on the seal tube flange inside the tunnel
- Quick release sealing at the seal tube bellmouth (to be removed by ROV prior to pipeline pull-in)
- Hydraulic jack assembly for pipeline alignment and internal end cap for pipeline pullhead.
- Sufficient strength to withstand net riser/pipeline forces

The seals are installed during the tie-in prior to the pull-in operation. This allows for flexibility since one sealtube may accommodate a range in pipeline sizes. Henceforth, it is possible to pre-install tie-in arrangements suitable for future pipelines without knowing the future lines exact size. Further, it is possible to exchange an installed line with a line of different size.

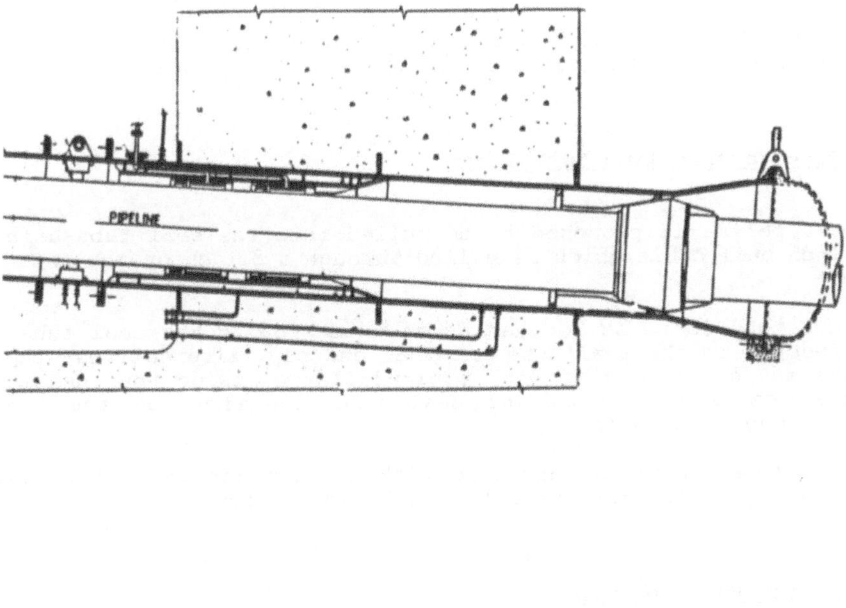

Figure 2 Sealtube

Seal Tube Sealing

Following the pipeline pull-in into the seal tube it will be necessary to seal the annulus between the pipeline and the seal tube.

Several alternative sealing systems may be applied:

- Inflatable seal system which can be installed into the seal tube prior to pipeline pull-in.
- Possible grouting of seal tube annulus as a back-up system
- A rubber plug seal to be attached to the pipeline and pulled into the seal tube bellmouth and thereby providing back-up sealing
- External sealing flange on the seal tube bellmouth

The seal tube has been designed to accommodate all the alternatives mentioned above. However, some of the systems are designed as back-up.

J-Tube for Pipeline Pull Cable

The pipeline is proposed to be pulled into the seal tube using a 4 inch pull cable which is pulled through a 8 inch preinstalled J-tube.

The J-tube is to be located as an extension to the seal tube and connected to the seal tube internal end cap. Alternatively the J-tube may be terminated with a blind flange inside the tunnel and later be connected to the seal tube depending on the tie-in technique to be used

The J-tube is to be installed with a corrosion inhibitor system which may be combined with the seal tube system.

Concrete Pull-In Ramp

In order to lead the pipeline into the seal tube bellmouth, it is proposed that a concrete ramp is to be installed at each seal tube entrance. The ramp is to be inclined at approximately 20° and projecting 9 m from the cell wall.

TIE-IN PROCEDURE

The following gives a sequential description of the tie-in procedure.

Figure 3 shows the overall layout of the tie-in arrangement.

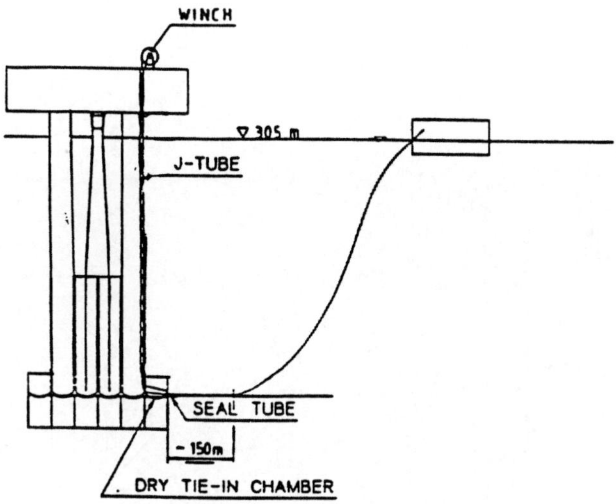

Figure 3 Tie-In Arrangement

1. Preinstalled Items

The following items are to be preinstalled:

- Riser
- Seal Tube (Without seal system)
- Tagline in seal tube and J-tube
- Inhibitor piping
- Drain/test valves
- Piping for seal tube/J-tube flooding
- Padeyes
- Auxiliary systems

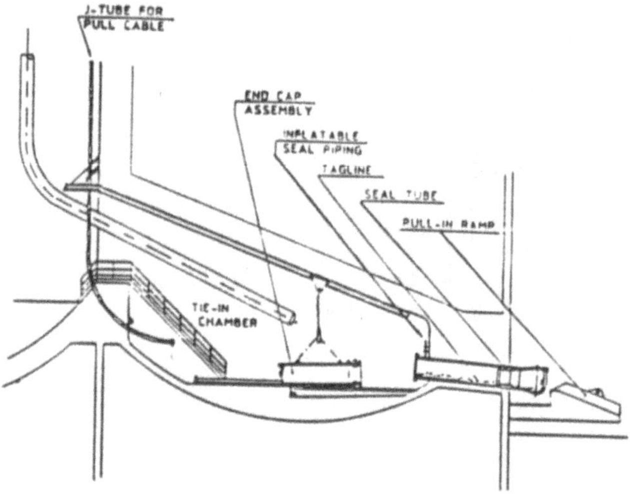

Figure 4

2. Preparation For Pull-In

The following work tasks need to be performed before pull-in is started:

- Install winch at topside.
- Drain inhibitor from seal tube and J-tube.
- Install seal system.
- Install end cap/hydraulic jack assembly, J-tube straight section and connect tagline.
- Flood seal tube and J-tube in predetermined sequence and check flange connections continuously.
- ROV connect a tagline to external end cap and disconnect the end cap.
- External end cap is pulled to surface by RSV.
- Connection of 4" pull cable to pipeline on the layvessel is commenced.

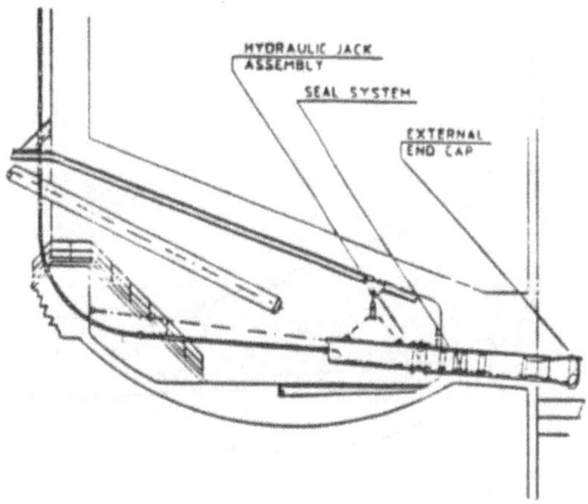

Figure 5

3. Pull-In And Sealing

The following tasks to be performed during the pull-in and sealing of the pipelines.

- Layvessel is to be located approximately 1000 metres from the platform.
- Laying and pulling of the pipeline is commenced.
- Pipeline is pulled to the entrance of the seal tube where the pulling is stopped.
- Pulling operation is being monitored by ROV.
- Waterjetting is performed on rubber plug seal and along pipeline to remove seabed particles.
- Pulling operation is carefully continued until pipeline is touching the internal end cap wall.
- Activate inflatable seals.
- Drain water from J-tube and end cap and test sealing.
- Option: Drain water from seal tube annulus in front of rubber plug.
- After careful testing of the sealing, preparation for end cap removal can start.

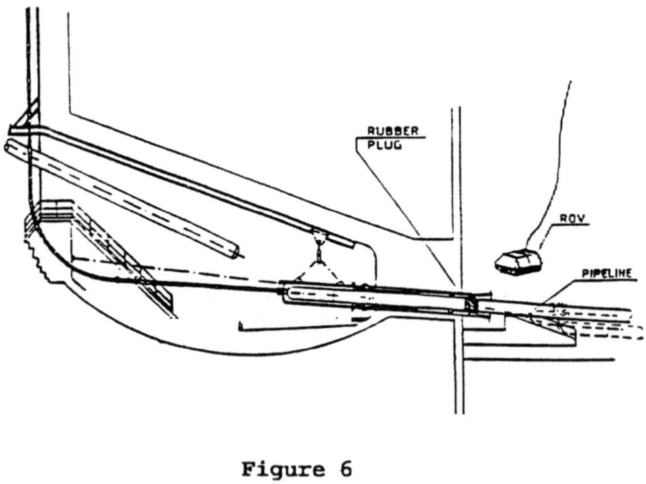

Figure 6

4. End Cap Removal And Installation Of Pipeline Anchor

The following tasks are to be performed:

- Pipeline is aligned with hydraulic jacks.
- End cap assembly is unbolted from hydraulic jack assembly and cut open at the end connected to the J-tube (Pull cable must be protected by fire proof socket).
- End cap assembly is moved towards the J-tube and cut in two halves while tension in the pipeline is maintained.
- Temporary hold-back clamp and pipeline support ring is installed on the pipeline.
- Pipeline pull head is removed.
- Pipeline anchor is installed.
- Split sealing spool is installed and temporary clamp removed.
- Seals and hydraulic jacks are released.

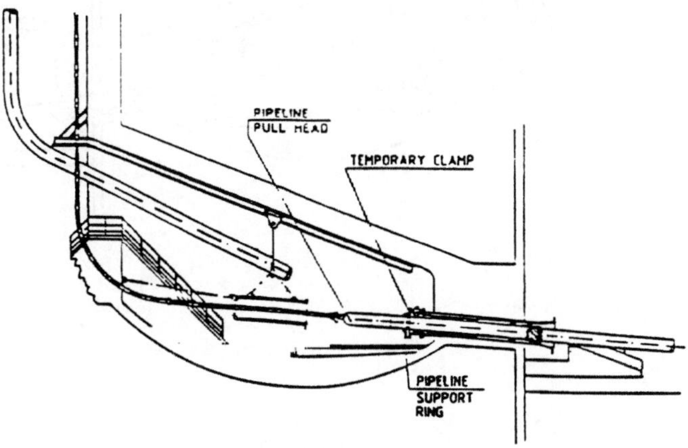

Figure 7

5. Riser Pipeline Spool

The following tasks ends the tie-in sequence:

- Spool is welded to riser and pipeline.
- Clean up.
- Laying of pipeline is commencing with a radius of approximately 1500 metres.
- External rubber seal can be installed at the bellmouth and seal tube refilled with inhibitor for corrosion protection.

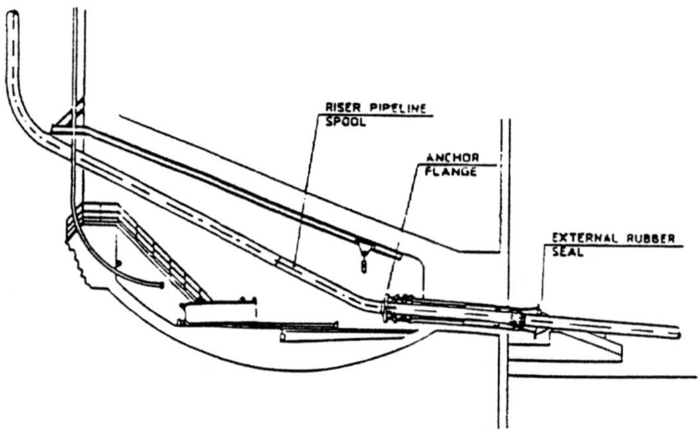

Figure 8

SCHEDULE

The following shows a schedule for the tie-in of the pipeline to the platform. The schedule shows the following main activities:

- Preparation work, 10 days.
- Operations estimated to tie-in, 14 days.
- Pipelaying

	Preparation Activities	10 days	7 days	7 days	
1.	Preparation work. Remov cap, est.-pull-in wire, survey, etc.				
2.0	Pipelaying				
2.1	Mob/Sail				
2.2	Positioning				
2.3	Pick-Up cable				
2.4	Pay out pipel.				
2.5	Pull-in				
2.6	Alignment, finalize				
2.7	Cont. pipe-lying - 1km				
2.8	Act. seals				
2.9	Cont. pipe-laying 2 km				
2.10	Test Seals dewater				
2.11	Remove int-ernal end cap				
2.12	Install temp. clamp				
2.13	Riser Tie-in				
2.14	Cont. pipe-laying				

COMPUTER ANALYSIS

In order to establish design loads on the sealtube, various typical computer runs have been carried out using the non-linear finite element program ABAQUS.

Three pipeline layouts were investigated:

1) Pipeline approaching sealtube in a straight line.

2) Pipeline approaching pipeline with a radius of 1500 metres.

3) Pipeline approaching sealtube with a 80 metre long dogleg.

The following loads were considered in the analyses:

- Deadweight
- Differential pressure
- Temperature
- Relative platform settlement
- Platform horizontal movement due to wave loading.

As a restraining force, seabed friction was also considered.

Two types of pipeline models were made, one global and one local. The global model was used to establish forces and displacements which were then incorporated as boundary conditions for the local model.

The following discussions apply to the analyses made for a 40" pipeline.

Straight Pipeline

The seal tube anchor will be subjected to an axial compressive load of 1580 tons and 1620 tons in the functional and functional + environmental load case respectively. This since the full end cap load will be applied on the sealtube together with loads arising from temperature expansion. These high compressive loads, linked with deflections, cause large bending stresses around the entrance in the pipeline. The stresses were found to be 530 and 590 MPa for the two load conditions respectively.

Curved Pipeline

For the curved pipeline approach with a radius of 1500 metres the pipeline was found to be in tension at all times. This since the curve neutralises the end cap loading. However, due to the differential pressure between the inside of the shaft and the outside at 300 metres water depth, a net tension loading is introduced.

The seal tube anchor is subjected to a net axial tension load of 220 tons and 360 tons in the functional and functional + enviromental load case respectively. The derived tension loads includes thermal expansion effects. Since the tension loads reduce the bending stresses around the entrance, the corresponding stresses are low; 200 and 220 MPa for each of the two load conditions.

Dogleg

A dogleg approach gives rise to net axial tension loads of 150 tons and 160 tons in the functional and functional + environmental loadcase, respectively. The corresponding stresses in the pipeline were found to be 240 MPa for both loading conditions. The axial deflection in the dogleg bend due to pipeline expansion was 2.9 metres.

Concluding Remarks on Analysis

The analyses shows that the curved pipeline and dogleg solutions give acceptable design loads and stresses in the 40" pipeline. However, the straight pipeline approach causes loads and stresses that are too high for practical design purposes.

ALTERNATIVE TIE-IN METHODS

This section shows some alternative tie-in arrangements.

The alternative, shown in figure 9 and 10, is the same as the previously outlined tie-in concept, but uses one or two sheaves in stead of internal J-tube during the pull-in operation.

Using a subsea sheave will lead to increased subsea work with corresponding increased costs.

Figure 11 through 15 show alternative installation methods involving hyperbaric welding and diving. However, one shall bear in mind that if remotely operated and reliable hyperbaric welding systems can be used, these alternatives may well prove to be more cost effective than the outlined tie-in arrangement.

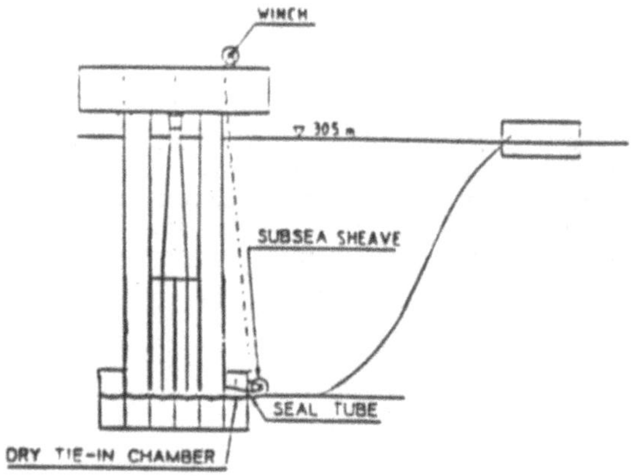

Figure 9

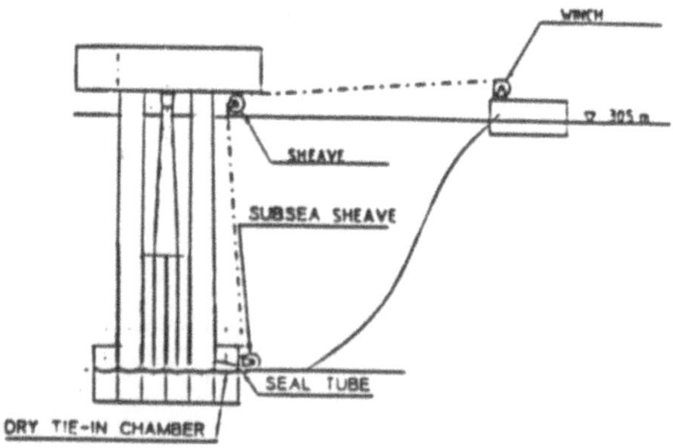

Figure 10

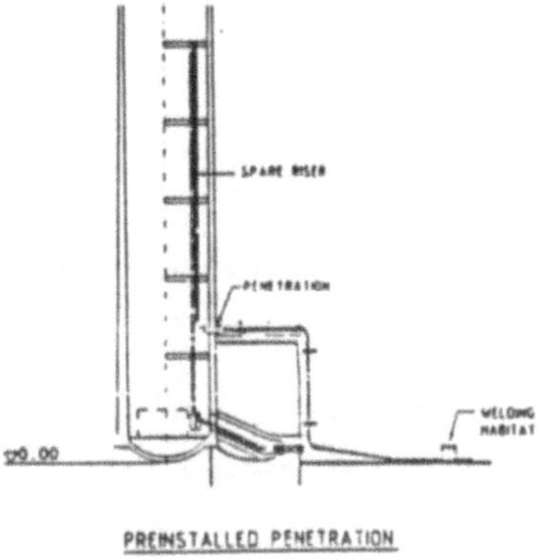

Figure 11

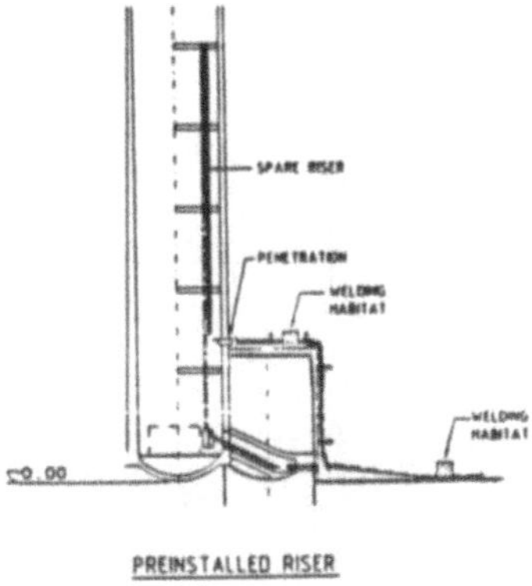

Figure 12

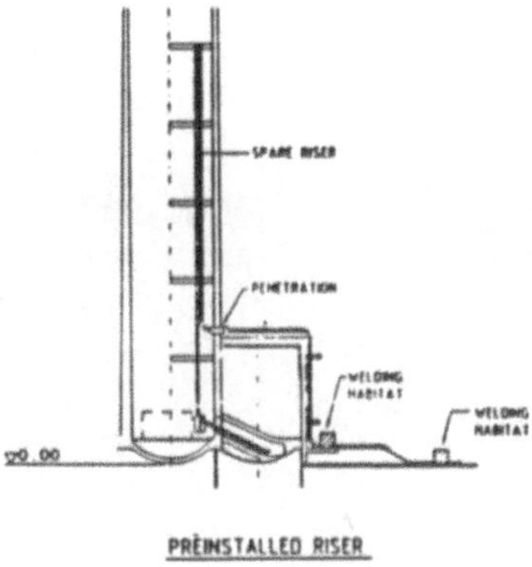

Figure 13

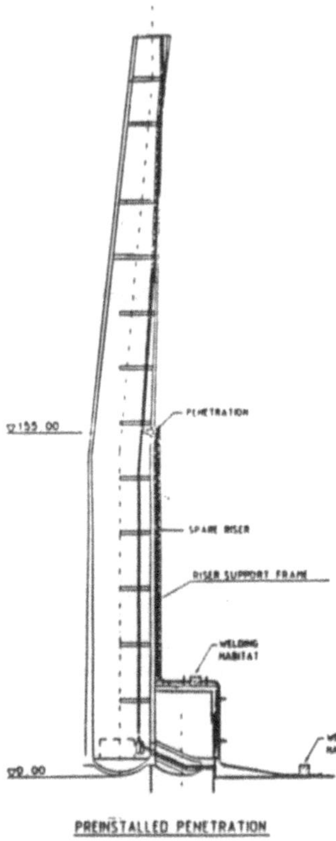

Figure 14

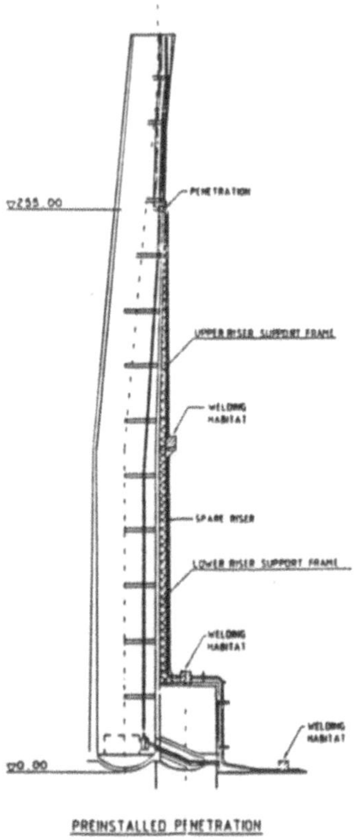

Figure 15

ACKNOWLEDGEMENT

The authors will use this oppertunity to thank INOCEAN Construction A/S for their assistance with regards to the feasibility of the tie-in concept.

INDEX

GPSR Compliance

The European Union's (EU) General Product Safety Regulation (GPSR) is a set of rules that requires consumer products to be safe and our obligations to ensure this.

If you have any concerns about our products, you can contact us on ProductSafety@springernature.com

In case Publisher is established outside the EU, the EU authorized representative is:

Springer Nature Customer Service Center GmbH
Europaplatz 3
69115 Heidelberg, Germany

Batch number: 10607220

Printed by Printforce, the Netherlands